Ecological Research Monographs

Series Editor

Yoh Iwasa, Kyushu University, Fukuoka, Japan

The book series Ecological Research Monographs publishes refereed volumes on all aspects of ecology, including Animal ecology Population ecology Theoretical ecology Plant ecology Community ecology Statistical ecology Marine ecology Ecosystems Biodiversity Microbial ecology Landscape ecology Conservation Molecular ecology Behavioral ecology Urban ecology Physiological ecology Evolutionary ecology The series comprise books and edited collections by international experts in their fields.

Toshiaki Owari • Satoshi N. Suzuki
Nobuaki Tanaka • Dai Fukui
Editors

Integrative Forest Management and Silviculture

Harmonizing Conservation and Production

Springer

Editors
Toshiaki Owari
The University of Tokyo Hokkaido Forest, Graduate School of Agricultural and Life Sciences
The University of Tokyo
Furano, Japan

Nobuaki Tanaka
The University of Tokyo Hokkaido Forest, Graduate School of Agricultural and Life Sciences
The University of Tokyo
Furano, Japan

Satoshi N. Suzuki
Nakagawa Experimental Forest, Field Science Center for Northern Biosphere
Hokkaido University
Otoineppu, Japan

Dai Fukui
The University of Tokyo Fuji Iyashinomori Woodland Study Center, Graduate School of Agricultural and Life Sciences
The University of Tokyo
Yamanakako, Japan

ISSN 2191-0707 ISSN 2191-0715 (electronic)
Ecological Research Monographs
ISBN 978-981-95-1016-0 ISBN 978-981-95-1017-7 (eBook)
https://doi.org/10.1007/978-981-95-1017-7

Cover photo: Conifer–broadleaf mixed forest with six selection cuttings over 60 years.

This Springer imprint is published by the registered company Springer Nature Singapore Pte Ltd.
The registered company address is: 152 Beach Road, #21-01/04 Gateway East, Singapore 189721, Singapore

Preface

As climate change increasingly threatens the effectiveness of traditional forestry practices, sustainable forest management must evolve to prioritize resilience, biodiversity, and ecosystem functions. This book presents the integrative forest management and silvicultural practices developed by the University of Tokyo Hokkaido Forest (UTHF) over the past six decades. Central to the UTHF's approach is the Stand-based Silvicultural Management System (SSMS), a pioneering framework designed to balance timber production with biodiversity conservation and ecosystem health. This book provides an in-depth exploration of the SSMS and its application at the UTHF, serving as a model for integrative forest management applicable to diverse forest ecosystems worldwide.

The book explores the ecological, climatic, and geological dynamics that shape the UTHF′s diverse hemi-boreal forest ecosystem, as well as the history and evolution of forest management at the UTHF, including its responses to large-scale disturbances and its integration of cutting-edge digital technologies. It outlines the six foundational principles of the SSMS, which guide practices to maintain multi-storied stands, retain trees with superior traits, and foster mixed conifer-broadleaf forests, along with specific silvicultural practices, such as single-tree selection harvesting, assisted natural regeneration, and the production of high-quality seedlings, all of which are essential for ensuring the forest's long-term health and productivity. Additionally, the rich biodiversity of the UTHF, supported by extensive floral and faunal survey data, and the conservation measures implemented to protect this biodiversity. It also discusses the UTHF′s contributions to vital ecosystem services, including carbon sequestration, water supply, and cultural heritage.

Through case studies and detailed analysis, this book serves as a valuable resource for forestry scientists, practitioners, and policymakers. It offers insights into sustainable forest management in a changing world. By sharing the experiences and lessons learned from the UTHF, the book aims to inspire and guide the advancement of sustainable forestry practices globally.

Furano, Japan — Toshiaki Owari
Otoineppu, Japan — Satoshi N. Suzuki
Furano, Japan — Nobuaki Tanaka
Yamanakako, Japan — Dai Fukui

Species Nomenclature and Usage in This Book

Latin Names

The Latin names of plant species in this book are primarily based on Horie et al. (2013), although some of them are not accepted names, but rather synonyms, in the global Latin name database, the WFO Plant List (https://wfoplantlist.org/).

Name in Horie et al. 2013	Accepted name in WFO Plant List
Quercus crispula	*Quercus mongolica* var. *crispula*
Padus ssiori	*Prunus ssiori*
Cortusa matthioli subsp. *pekinensis*	*Primula matthioli* subsp. *pekinensis*
Betula platyphylla var. *japonica*	*Betula pendula* subsp. *mandshurica*

English Names

In some chapters, English common names are majorly used instead of Latin names to enhance readability for non-academic practitioners.

The tree species for which English name is used are as follows:

Sakhalin fir: *Abies sachalinensis*
Yezo spruce: *Picea jezoensis*
Sakhalin spruce: *Picea glehnii*
Japanese oak: *Quercus crispula*
Japanese elm: *Ulmus davidiana* var. *japonica*
Manchurian elm: *Ulmus laciniata*
Japanese linden: *Tilia japonica*
Maximowicz's lime: *Tilia maximowicziana*
Painted maple: *Acer pictum*
Japanese white birch: *Betula platyphylla* var. *japonica*

Elman's birch: *Betula ermanii*
Monarch birch: *Betula maximowicziana*
Manchurian alder: *Alnus hirsuta* var. *hirsuta*
Katsura tree: *Cercidiphyllum japonicum*
Castor aralia: *Kalopanax septemlobus*
Manchurian ash: *Fraxinus mandshurica*
Japanese ash: *Fraxinus lanuginose*
Hokkaido bird cherry: *Padus ssiori*
Japanese big-leaf magnolia: *Magnolia obovata*
Goat willow: *Salix caprea*
Japanese larch: *Larix kaempferi*
Kurile larch: *Larix gmelinii* var. *japonica*
Korean larch: *Larix gmelinii* var. *olgensis*
Siberian larch: *Larix sibirica*
American larch: *Larix laricina*
Eastern white pine: *Pinus strobus*
Norway spruce: *Picea abies*
Scots pine: *Pinus sylvestris*
Jack pine: *Pinus banksiana*

Reference

Horie K, Miyamoto Y, Kimura N, Oikawa N (2013) List of vascular plant of the University of Tokyo Hokkaido Forest, Miscellaneous Information of The University of Tokyo Forests 54:59–106.

Acknowledgments

This book was published to commemorate the 125th anniversary of the University of Tokyo Hokkaido Forest. We extend our heartfelt gratitude to everyone who supported and encouraged us during the preparation of this publication.

We are profoundly grateful to all the staff who have contributed to the University of Tokyo Hokkaido Forest over the past 125 years. Their dedication, expertise, and unwavering commitment have been instrumental in shaping the forest's legacy and enabling this publication. Their work has laid a strong foundation for the research and conservation efforts that continue to flourish today.

We would also like to acknowledge the use of AI-based tools, including Curie (https://www.aje.com/curie/), DeepL (https://www.deepl.com/), ChatGPT (OpenAI, https://openai.com/chatgpt/), and NotebookLM (https://notebooklm.google.com/), which provided valuable assistance in proofreading and enhancing the clarity of the text.

Our special thanks to Springer Nature for their trust and professionalism in bringing this book to fruition, and to the editorial team for their meticulous attention to detail.

This publication was made possible through the generous contributions to the University of Tokyo Foundation's "The University of Tokyo Hokkaido Forest 125th Anniversary Support Fund." We extend our deepest gratitude to all the donors for their invaluable support.

Finally, we are deeply indebted to the many individuals who generously shared their experiences and knowledge, offering inspiration and perspective. Their contributions have been indispensable in making this book a reality.

To all who have been part of this milestone for the University of Tokyo Hokkaido Forest, we express our deepest gratitude.

Contents

Contributors

Dai Fukui The University of Tokyo Fuji Iyashinomori Woodland Study Center, Graduate School of Agricultural and Life Sciences, The University of Tokyo, Yamanakako, Japan

Susumu Goto The University of Tokyo Tanashi Forest, Graduate School of Agricultural and Life Sciences, The University of Tokyo, Nishitokyo, Japan

Naoto Kamata The University of Tokyo Chiba Forest, Graduate School of Agricultural and Life Sciences, The University of Tokyo, Kamogawa, Japan

Toshiaki Owari The University of Tokyo Hokkaido Forest, Graduate School of Agricultural and Life Sciences, The University of Tokyo, Furano, Japan

Daisuke Sakaue The University of Tokyo Chichibu Forest, Graduate School of Agricultural and Life Sciences, The University of Tokyo, Chichibu, Japan

Satoshi N. Suzuki The University of Tokyo Hokkaido Forest, Graduate School of Agricultural and Life Sciences, The University of Tokyo, Furano, Japan

Nakagawa Experimental Forest, Field Science Center for Northern Biosphere, Hokkaido University, Otoineppu, Japan

Nobuaki Tanaka The University of Tokyo Hokkaido Forest, Graduate School of Agricultural and Life Sciences, The University of Tokyo, Furano, Japan

Abbreviations

≡	Homotypic synonym
=	Heterotypic synonym
a.s.l.	Above sea level
AIST	National Institute of Advanced Industrial Science and Technology
As	*Abies sachalinensis*
CA	Community-based association
CS	Curvature and slope
DBH	Diameter at breast height
ETH	Swiss Federal Institute of Technology
GIS	Geographic information system
GNSS	Global navigation satellite system
GPS	Global positioning system
ha	Hectare
JAS	Japanese Agricultural Standards
JMA	Japan Meteorological Agency
JPY	Japanese yen
LiDAR	Light detection and ranging (laser scanning)
LTER	Long Term Ecological Research
NDVI	Normalized difference vegetation index
NFWG	Natural Forest Working Group
Pg	*Picea glehnii*
Pj	*Picea jezoensis*
RTK	Real-time kinematic
SCWG	Selective Cutting Working Group
SS	Simplified water purification station
SSMS	Stand-based Silvicultural Management System
UAV	Unmanned aerial vehicle
UTCBF	The University of Tokyo Chiba Forest
UTF	The University of Tokyo Forests
UTHF	The University of Tokyo Hokkaido Forest

Part I
Introduction

Chapter 1
Outline of This Book

Toshiaki Owari, Satoshi N. Suzuki, Nobuaki Tanaka, and Dai Fukui

Abstract As climate change increases the vulnerability of traditional silvicultural systems, sustainable forestry practices must evolve to prioritize resilience, biodiversity, and ecosystem functions. This book explores the integrative forest management and silvicultural practices developed and applied at the University of Tokyo Hokkaido Forest (UTHF) over the past 60 years. The Stand-based Silvicultural Management System (SSMS) is a pioneering approach that balances timber production, biodiversity conservation, and ecosystem health. The book examines the ecological, climatic, and geological dynamics shaping the UTHF's complex hemi-boreal forest ecosystem. It highlights adaptive silvicultural practices, such as single-tree selection harvesting and assisted natural regeneration, which maintain forest diversity and stability. The SSMS's evolution is supported by long-term ecological research and advanced digital technologies, enabling precise forest classification and sustainable resource management. Case studies on recovery from large-scale disturbances, biodiversity conservation, and ecosystem service optimization illustrate how the UTHF integrates ecological, social, and cultural objectives. Challenges such as invasive species, deer overpopulation, and climate-induced disturbances underscore the need for adaptive strategies, including transitioning from monoculture plantations to mixed-species forests. By sharing insights into the SSMS, this

T. Owari (✉) · N. Tanaka
The University of Tokyo Hokkaido Forest, Graduate School of Agricultural and Life Sciences, The University of Tokyo, Furano, Japan
e-mail: owari@g.ecc.u-tokyo.ac.jp; tnk-nobu@g.ecc.u-tokyo.ac.jp

S. N. Suzuki
The University of Tokyo Hokkaido Forest, Graduate School of Agricultural and Life Sciences, The University of Tokyo, Furano, Japan

Nakagawa Experimental Forest, Field Science Center for Northern Biosphere, The University of Tokyo, Otoineppu, Japan
e-mail: snsuzuki@fsc.hokudai.ac.jp

D. Fukui
The University of Tokyo Fuji Iyashinomori Woodland Study Center, Graduate School of Agricultural and Life Sciences, The University of Tokyo, Yamanakako, Japan
e-mail: fukuidai@g.ecc.u-tokyo.ac.jp

T. Owari et al. (eds.), *Integrative Forest Management and Silviculture*,
Ecological Research Monographs, https://doi.org/10.1007/978-981-95-1017-7_1

book provides a valuable resource for advancing forestry science and sustainable forest management globally, offering a model for integrative practices in a changing world.

Keywords Ecosystem function · Forest ecosystem · Forest management · Silviculture · The University of Tokyo Hokkaido Forest

1.1 Introduction

As climate change intensifies, traditional silvicultural systems—characterized by clearcutting and even-aged monoculture plantations—are becoming increasingly vulnerable to disturbances such as windstorms and insect outbreaks. These methods simplify forest ecosystems, reducing their complexity and degrading their resilience. Modern forestry must therefore prioritize not only sustainable timber production but also the preservation of key ecosystem functions, including biodiversity conservation, water and nutrient cycling, and the provision of cultural services.

Since the early 1990s, alternative forest management approaches have gained global attention. These approaches emphasize the integration of natural processes into forest management to maintain structural and compositional diversity, thereby enhancing ecosystem services and improving resilience to climate change (Puettmann et al. 2015; Bauhus et al. 2013). Examples include close-to-nature forestry (Brang et al. 2014; Larsen et al. 2022; O'Hara 2016), retention forestry (Gustafsson et al. 2012; Lindenmayer et al. 2012), and ecological forestry (Franklin 1989; Palik et al. 2020).

Integrative forest management represents a promising alternative, balancing forest production, conservation, and ecosystem protection (Kraus and Krumm 2013). Unlike the "segregation" approach, which assigns specific objectives to separate forest zones (Bončina 2011), integrative management pursues multiple goals within the same area. These include timber and nontimber production, recreation, and biodiversity conservation (Bončina and Čavlović 2009; Bončina 2011). Typically, this approach employs low-impact logging techniques over large areas, minimizing the need for designated protected zones while supporting native biodiversity (Nagel et al. 2017). Also known as integrated forest management, this method has been implemented at the enterprise level in Europe (Krumm et al. 2021).

A pioneering example of integrative management is the Stand-based Silvicultural Management System (SSMS), developed by the University of Tokyo Hokkaido Forest (UTHF) in 1958. Designed to manage the UTHF's 23,000 hectares of forest in northern Japan, the SSMS adapts management practices to the specific conditions of each stand, balancing economic and ecological functions. With over six decades of application, the SSMS exemplifies sustainable integrative forest management in practice.

This book examines the integrative forest management and silvicultural practices employed by the UTHF over the past 60 years. Our goal is to share insights into the

SSMS, providing a valuable resource for advancing global forestry and forest science. Additionally, we aim to offer a practical reference for integrative forest management and silviculture in Japan and internationally.

This book is divided into five parts. Part I includes this outline and Chap. 2, which traces the history of the UTHF's forest management since 1899. Part II focuses on the UTHF's forest ecosystems, laying the groundwork for nature-based silviculture. Part III explores the SSMS as a system of integrative forest management and silviculture. Part IV examines the management and conservation of multiple ecosystem functions. Part V addresses future perspectives, discussing challenges and opportunities for advancing integrative forest management in a rapidly changing environment.

Chapter 2 in Part I highlights the UTHF's forest management history, from early colonization and afforestation efforts following fire damage to the current focus on sustainability and integration. This chapter offers valuable insights into the long-term application of integrative forest management principles by analyzing the development and outcomes of the UTHF's management strategies.

1.2 Forest Ecosystem

The SSMS adopts a nature-based, integrative approach that leverages ecosystem processes. Effective implementation requires a comprehensive understanding of forest ecosystem structure and dynamics, as well as the climatic and geological factors that influence them. Part II of this book focuses on the forest ecosystem and geographical characteristics of the UTHF, where the SSMS is applied.

Chapter 3 provides an overview of the UTHF's geographical and geological characteristics, emphasizing its diverse topography, soil types, and environmental complexity. This chapter lays the groundwork for understanding how these factors influence forest management and ecosystem processes. Chapter 4 examines the climate and hydrology of the UTHF, with a focus on its subarctic climate, long-term meteorological trends, and seasonal variations in precipitation and snowfall. It also explores the forest's hydrological features, including streamflow dynamics, groundwater contributions, and the influence of geological factors on water chemistry and flow patterns. Chapter 5 delves into the forest ecosystems of the UTHF, highlighting vegetation patterns along elevation gradients and the impacts of natural disturbances. Long-term ecological studies provide insights into biodiversity, forest structure, and responses to climate change, informing sustainable management practices. Chapter 6 highlights the rich biodiversity of the UTHF, presenting detailed findings from faunal surveys of insects, mammals, birds, and fish. This chapter underscores the importance of understanding species interactions and environmental factors for effective conservation and forest management.

1.3 Forest Management and Silviculture

Part III explores the fundamental principles and practical applications of the SSMS, a forest management system implemented at the UTHF since 1958. This part examines how the SSMS integrates ecological principles into forestry practices, balancing timber production with environmental conservation.

Chapter 7 introduces the overall concept of the SSMS, outlining its principles of adaptive management, minimal intervention, and the promotion of multilayered forests. This integrative approach seeks to harmonize timber production with environmental conservation. Chapter 8 outlines the planning and implementation of silvicultural practices under the SSMS, including forest type classification, single-tree selection harvesting, and advanced survey techniques. It emphasizes sustainable timber production while maintaining forest health and biodiversity. Chapter 9 focuses on timber harvesting, marketing, and forest road development at the UTHF, highlighting innovations in mechanized logging and sustainable marketing strategies. The chapter also emphasizes the importance of maintaining infrastructure to support these operations. Chapter 10 reviews regeneration strategies, including techniques for assisting natural regeneration and plantation techniques to address challenges such as dwarf bamboo dominance. It traces the evolution of these strategies, particularly the shift from monoculture plantations to polyculture systems aimed at enhancing resilience and biodiversity. Chapter 11 addresses the production of high-quality seedlings to support conifer regeneration within the SSMS framework. It highlights the sourcing of seeds from clonal seed orchards and seed stands in the UTHF, production methods for bare-rooted and containerized seedlings, and challenges such as diseases and climate-related damage.

1.4 Management and Conservation of Multiple Ecosystem Functions

Part IV shifts focus to another key objective of the SSMS: optimizing the multiple ecosystem functions and services provided by forests at the landscape scale. This section illustrates how the UTHF integrates ecological, social, and cultural objectives into its forest management practices, positioning itself as a model for integrative forestry in an era of rapid environmental change.

Chapter 12 explores the ecological impacts of various forest management practices at the UTHF. It examines how these practices affect ecosystem processes, focusing on the reduction of dead wood volume and its effects on tree regeneration and biodiversity. The chapter also addresses challenges such as changes in forest structure, bark beetle infestations, and deer overgrazing, emphasizing the need for sustainable strategies that balance timber production with biodiversity conservation and ecosystem health. Chapter 13 examines forest recovery at the UTHF following large-scale disturbances like typhoons and wildfires. It analyzes recovery strategies, including salvage logging and reforestation, and discusses their long-term impacts on carbon storage, biodiversity, and resilience to climate change. Chapter 14 explores the effects of pests, diseases, and mammals on forest health and productivity. It highlights outbreaks, patterns of vulnerability, and the importance of monitoring and management to maintain forest ecosystem stability and timber production. Chapter 15 addresses biodiversity conservation at the UTHF, focusing on the SSMS framework and challenges such as deer overpopulation, climate change, and invasive species. It underscores the need to balance sustainable forestry with urgent conservation efforts. Chapter 16 evaluates the ecosystem services provided by the UTHF, with particular emphasis on its role in carbon sequestration and broader environmental contributions. It details methods for estimating carbon stocks and highlights the forest's ability to offset local carbon emissions. Additionally, it examines the UTHF's role in providing water for drinking and irrigation and supporting cultural activity of local communities.

1.5 The Future

Chapter 17 in Part V discusses the challenges and opportunities for future research and management at the UTHF. The SSMS is still evolving and is not yet a perfect forestry system. Ongoing efforts, through trial and error, aim to address various issues. This system must be further refined by integrating new technologies and adapting to environmental changes, such as climate change and deer overpopulation. This chapter outlines a pathway toward an integrative approach to forest management that balances ecosystem services, including timber production, carbon storage, and biodiversity conservation.

References

Bauhus J, Puettmann KJ, Kuhne C (2013) Close-to-nature forest management in Europe: does it support complexity and adaptability of forest ecosystem? In: Messier C, Puettmann KJ, Coates KD (eds) Managing forests as complex adaptive systems: building resilience to the challenge of global change. Routledge, New York, pp 187–213

Bončina A (2011) Conceptual approaches to integrate nature conservation into forest management: a central European perspective. Int For Rev 13(1):13–22

Bončina A, Čavlović J (2009) Perspectives of forest management planning: Slovenian and Croatian experience. Croat J For Eng 30(1):77–87

Brang P, Spathelf P, Larsen JB, Bauhus J, Andrej B, Chauvin C, Drössler L, García-Güemes C, Heiri C, Kerr G, Lexer MJ, Mason B, Mohren F, Mühlethaler U, Nocentini S, Svoboda M (2014) Suitability of close-to-nature silviculture for adapting temperate European forests to climate change. Forestry 87(4):492–503

Franklin JF (1989) Toward a new forestry. Am For 95:37–44

Gustafsson L, Baker SC, Bauhus J, Beese WJ, Brodie A, Kouki J, Lindenmayer DB, Lõhmus A, Pastur GM, Messier C, Neyland M, Palik B, Sverdrup-Thygeson A, Volney WJA, Wayne A,

Franklin JF (2012) Retention forestry to maintain multifunctional forests: a world perspective. BioScience 62(7):633–645
Kraus D, Krumm F (2013) Integrative approaches as an opportunity for the conservation of forest biodiversity. 284 pp. European Forest Institute
Krumm F, Schuck A, Rigling A (2021) How to balance forestry and biodiversity conservation? – a view across Europe. 640 pp. European Forest Institute, Swiss Federal Institute for Forest, Snow and Landscape Research, Birmensdorf
Larsen JB, Angelstam P, Bauhus J, Carvalho JF, Diaci J, Dobrowolska D, Gazda A, Gustafsson L, Krumm F, Knoke T, Konczal A, Kuuluvainen T, Mason B, Motta R, Pötzelsberger E, Rigling A, Schuck A (2022) Closer-to-nature forest management, From science to policy 12. European Forest Institute
Lindenmayer DB, Franklin JF, Lõhmus A, Baker SC, Bauhus J, Beese W, Brodie A, Kiehl B, Kouki J, Pastur GM, Messier C, Neyland M, Palik B, Sverdrup-Thygeson A, Volney J, Wayne A, Gustafsson L (2012) A major shift to the retention approach for forestry can help resolve some global forest sustainability issues. Conserv Lett 5(6):421–431
Nagel TA, Firm D, Pisek R, Mihelic T, Hladnik D, de Groot M, Rozenbergar D (2017) Evaluating the influence of integrative forest management on old-growth habitat structures in a temperate forest region. Biol Conserv 216:101–107
O'Hara KL (2016) What is close-to-nature silviculture in a changing world? Forestry 89(1):1–6
Palik BJ, D'Amato AW, Franklin JF, Johnson KN (2020) Ecological silviculture: foundations and applications. Waveland Press
Puettmann KJ, Wilson SM, Baker SC, Donoso PJ, Drössler L, Amente G, Harvey BD, Knoke T, Lu Y, Nocentini S, Putz FE, Yoshida T, Bauhus J (2015) Silvicultural alternatives to conventional even-aged forest management – what limits global adoption? For Ecosyst 2(1):8

Chapter 2
Management History of the University of Tokyo Hokkaido Forest

Toshiaki Owari and Satoshi N. Suzuki

Abstract This chapter provides a brief history of the management of the University of Tokyo Hokkaido Forest (UTHF), northern Japan. In 1899, Tokyo Imperial University acquired 23,597 ha of virgin forest from the Hokkaido Agency to serve as an experimental site for the College of Agriculture. Forest colonization, which included converting undeveloped forest lands into cultivated areas, settling immigrants, and involving them in both forestry work and pioneering tasks, was crucial to the early development of the UTHF. In the 1910s, severe forest fires caused significant damage, and the affected areas were subsequently treated. During 1920s and 1930s, the UTHF focused on expanding its management base while scaling back its afforestation project. Forests situated in the countryside experienced overcutting due to the demand for timber for military use and firewood collection in areas for enhancing tree growth through silvicultural operations during and after the war. In 1958, the UTHF initiated a business-scale experiment of the "Stand-Based Silvicultural Management System." A typhoon struck the UTHF in 1981, damaging approximately 40% of the total forested area. After addressing the wind-damaged areas, further developments were made based on the prior outcomes of the SSMS. Since the 2010s, the UTHF has been leading the digitalization of natural forest management operations.

Keywords Digital transformation · Forest colonization · Natural disturbance · Stand-based silvicultural management system

T. Owari (✉)
The University of Tokyo Hokkaido Forest, Graduate School of Agricultural and Life Sciences, The University of Tokyo, Furano, Japan
e-mail: owari@g.ecc.u-tokyo.ac.jp

S. N. Suzuki
The University of Tokyo Hokkaido Forest, Graduate School of Agricultural and Life Sciences, The University of Tokyo, Furano, Japan

Nakagawa Experimental Forest, Field Science Center for Northern Biosphere, Hokkaido University, Otoineppu, Japan
e-mail: snsuzuki@fsc.hokudai.ac.jp

T. Owari et al. (eds.), *Integrative Forest Management and Silviculture*,
Ecological Research Monographs, https://doi.org/10.1007/978-981-95-1017-7_2

2.1 Establishment and Its Beginnings (1889–1910)

The University of Tokyo Hokkaido Forest (UTHF)[1] was established in 1899 as an experimental site of the College of Agriculture, Tokyo Imperial University, to serve as a research and education center for forestry and forest science in the northern region. Dairoku Kikuchi, the fifth President of the Tokyo Imperial University, along with Naokichi Matsui, Dean of the College of Agriculture, and Zentaro Kawase, Director of the University Forests, led the establishment process. The University acquired a large tract of virgin forest spanning 23,597 ha from the Hokkaido Agency, under the Ministry of Home Affairs. Seiroku Honda, who was an associate professor of the Department of Forestry at the time, played a crucial role in site selection and negotiation efforts (Miura 1923).

Kintaro Hondo, the first local staff member of the UTHF, arrived in Furano in July 1900. His mission was to establish the boundaries and to survey the forest (Tokyo University Forests 1950). By December 1901, a branch office had been established near Shimo-Furano Station (now JR Furano Station). In June 1905, Tsuneo Miura (Fig. 2.1), Research Assistant formerly with the Tokyo Imperial University Forest, Chiba, was transferred to serve as the Chief (later, Director of the regional forest) of the UTHF. Upon transferring from the Hokkaido Agency, an additional stipulation mandated that the forest be exchanged for other forests if later deemed necessary for the reclamation of Hokkaido (Takahashi and Takahashi 1944).

In September 1905, Chief Miura submitted an "Opinion on the Management of the University Forest, Hokkaido" to the University, proposing forest management under a forest colonization system (Tokyo University Forests 1950; Oohashi and Arinaga 2004). Forest colonization involved cultivating land within uncultivated forests, bringing in immigrants, and engaging them in both forest labor and pioneering activities (Takahashi and Takahashi 1944). Due to the small local population at that time and the challenge of finding forest workers, the forest colonization system played an important role at the UTHF (Tokyo University Forest in Hokkaido 1973). In the same year, the UTHF purchased several dozen kilograms of seeds of foreign tree species from the United States through the Yokohama Seed and Seedling Company, and a nursery was opened in Yamabe in the following year to begin seedling production trials (Tokyo University Forest in Hokkaido 1973). In 1906, small-scale cultivation was first attempted downstream of Onkozawa as a forest colonization experiment (Tokyo Imperial University Forest, Hokkaido 1934). The first 7705 m^3 of standing timber was sold in the same year (Tokyo University Forest in Hokkaido 1973).

[1]Originally, it was called as the Tokyo Imperial University Forest, Hokkaido (1899–1947), and later as the Tokyo University Forest in Hokkaido (1947–2015). Throughout this chapter, the abbreviation "UTHF" is used instead of its original name.

Fig. 2.1 Tsuneo Miura, the first Chief of the Tokyo Imperial University Forest, Hokkaido. (Photo by the UTHF, licensed under CC BY-NC-ND 4.0)

In 1907, *the 1st silviculture plan*[2] *(1907–1911)* was established along with regulations for leasing and renting land, marking the commencement of full-scale forest management. Implementing the plan was challenging due to numerous unknowns concerning boundaries, area, and forest conditions, making it impractical to manage the entire forest simultaneously (The University of Tokyo Forests 2022b). Consequently, the forest was divided into four subdivisions, with the most accessible 5000 ha (Compartments 8 and 10 in Fig. 2.2) selected for a 30-year consolidation period. In the old-growth forest, the average growing stock was approximately 300 m^3/ha. Many trees were decayed or bent, and the available stem volume was 100 m^3/ha for conifers and less than 20 m^3/ha for broad-leaved trees. The timber harvest policy focused on improving forest structure, primarily aiming to enhance areas with good regeneration potential. The planned annual tree harvest volume was 17,800 m^3. In addition to the sale of standing timber, the UTHF initiated the sale of logs and squared timber produced directly by the UTHF. This direct timber production gradually evolved into the primary sales approach (Miura 1923).

[2] Since the 1st plan, medium-term forest management plans at the UTHF have been formulated every 5–10 years, with naming conventions evolving over time. The 1st through 4th plans were called "segyo an" (silviculture plan), followed by "keiei an" (management plan) from the 5th to the 9th, "shiken kenkyu keikaku" (management and experiment plan) from the 10th to the 12th, and "kyoiku kenkyu keikaku" (education and research plan) from the 13th onward.

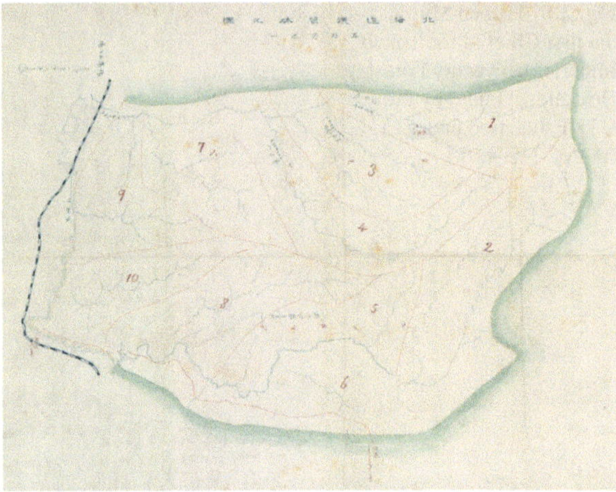

Fig. 2.2 Map of the Tokyo Imperial University Forest, Hokkaido, attached with the 1st silviculture plan (1907–1911). Red numbers and lines indicate the compartmet numbers and their boundaries, respectively. Blue lines indicate waterways. (Source: The University of Tokyo Hokkaido Forest)

2.2 Forest Damage Caused by Fires and Management of Fire-Damaged Sites (1911–1922)

In the spring of 1911, a severe forest fire, originating from the Model Forest owned by the Hokkaido Agency (i.e., forests transferred from the Japanese Government to set an example of forestry management and assist the Agency's finances), damaged 3600 ha of virgin forest (Fig. 2.3), 81 ha of planted forest, and approximately 240 thousand m^3 of standing trees over a period of 11 days (Tokyo University Forests 1950). The year was unusually dry, leading to large fires breaking out across various parts of Hokkaido (Tokyo University Forest in Hokkaido 1973). In 1914, another large fire occurred in the UTHF. After these fires, approximately 1500 ha of secondary forest was created through fire regeneration (The University of Tokyo Forests 2022b, also refer to Chap. 13).

Consequently, the first silviculture plan was terminated and **the 2nd silviculture plan (1911–1915)** was formulated, focusing on the treatment of fire-damaged trees (The University of Tokyo Forests 2022b). The new plan included an annual tree harvest volume of 69,000 m^3 and afforestation of 600 ha, approximately four times the volume of the first plan. Owing to its effectiveness in extinguishing fires and in order to secure labor, forest colonization was further promoted by the UTHF, resulting in 70 ha of land being cultivated. As a result, the damaged trees were cleared within five years, by 1915, with an annual tree harvest volume of 60,000 m^3. In the afforestation project, coniferous trees mostly with nonnative Norway spruce (*Picea abies*) were planted (Fig. 2.4) on 567 ha, which accounted for 95% of the planned area.

After completing the treatment of fire-damaged areas, **the 3rd silviculture plan (1916–1920)** was implemented. This plan focused on the management of fire-damaged sites, selective cutting in old-growth forests, and cultivation of agricultural lands. The entire forest, covering 24,012 ha, of which 414 ha were granted by the

Fig. 2.3 A forest damaged by fire in 1911. (Photo by the UTHF, licensed under CC BY-NC-ND 4.0)

Fig. 2.4 Norway spruce planted in 1911. (Photo taken in 1918 by the UTHF, licensed under CC BY-NC-ND 4.0)

Ministry of Home Affairs, was reorganized into one management unit comprising two subdistricts and 26 compartments (Tokyo University Forest in Hokkaido 2007; The University of Tokyo Forests 2022b). To replace old and overmature trees, address economic considerations, and manage constraints in logging and hauling operations, a rotation period of 25 years was established (Tokyo University Forest in Hokkaido 2007). Selective cutting served as the primary method in the old-growth forests, employing a cutting rate of 25–30%. Although the planned annual tree harvest volume was 38,200 m^3, the actual volume reached 74,000 m^3 due to the clear-cutting of 900 ha in the planned settlement area. The plan initially targeted afforestation of 1000 ha on the fire-damaged site; however, only 480 ha were actually planted, likely owing to good natural regeneration in the area (The University of Tokyo Forests 2022b).

In 1922 and 1923, the UTHF expanded its jurisdiction by purchasing 2533 ha of local government forest under the Hokkaido Agency, increasing its total area of to 26,545 ha (Tokyo University Forest in Hokkaido 2007). **The 4th silviculture plan (1921–1925)** was organized by expanding the number of compartments to 28, following the land acquisitions. Similar to the 3rd planning period, silvicultural operations included large-scale afforestation, clear-cutting in planned settlement areas, and selective cutting of old-growth forests. In the selective cutting areas, the rotation period was set at 25 years with a cutting rate of 25–30%. The annual tree harvest volume was planned to be 52,000 m^3, with 51,200 m^3 executed. Additionally, an average of 53,000 m^3 per year was harvested through clear-cutting of land designated for cultivation. Reforestation efforts in areas affected by forest fires, ongoing since the 3rd period, were nearly completed by around 1922. Species planted included Norway spruce, Scots pine (*Pinus sylvestris*), Japanese larch (*Larix kaempferi*), Eastern white pine (*Pinus strobus*), and European black pine (*Pinus nigra*), among others. However, severe damage caused by rodents resulted in only about one-third of the 1100 ha of forest planted with nonnative tree species being successfully established, with the remainder lost. Additionally, construction of forest tramways began in the 4th period (The University of Tokyo Forests 2022b).

2.3 Facility Expansion and Reduction of Afforestation (1923–1940)

Following Chief Miura's tenure (1905–1923), the second Chief, Kotaro Tomana (1923–1929), and the third Chief, Masaaki Nakayama (1929–1941), focused on expanding the management base while scaling back the afforestation project (Tokyo University Forest in Hokkaido 1973). Starting with **the 5th management plan (1926–1935, extended until 1937)**, the planning period was extended to 10 years but was further prolonged by two additional years due to the survey for revising forest compartment demarcation (The University of Tokyo Forests 2022b). As a result of the revised survey conducted in 1936–1937, the actual measurement increased by 3457 ha, and the total area of the UTHF was revised to 30,002 ha (Tokyo University Forest in Hokkaido 2007). The forest compartment system, which was reorganized into 26 compartments at the beginning of the 5th period (Fig. 2.5), was further reorganized into 109 compartments by its end (Fig. 2.6). The demarcation of these forest compartments is still in use today (The University of Tokyo Forests 2022b).

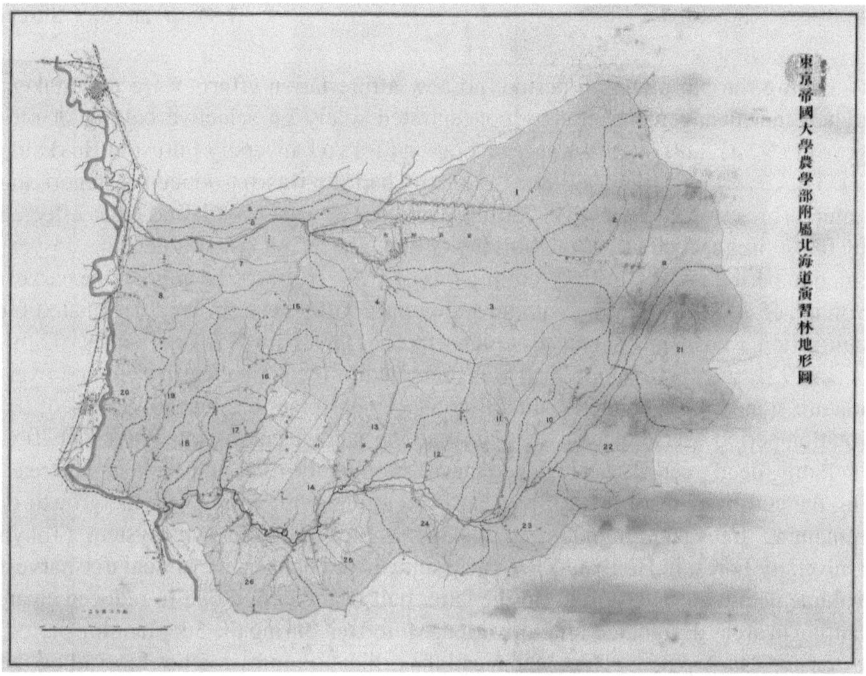

Fig. 2.5 Forest compartment map of the University of Tokyo Hokkaido Forest in 1931. Bold numbers and lines with dot indicate the compartment numbers and their boundaries, respectively. The red hatched areas show land developed for agriculture. (Source: The University of Tokyo Hokkaido Forest)

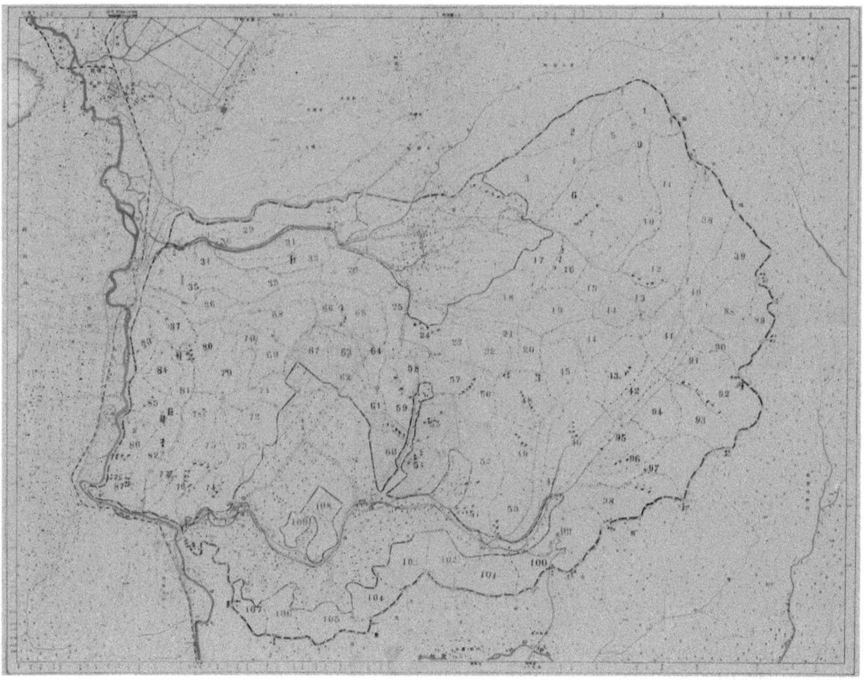

Fig. 2.6 The 109-comartment system of the University of Tokyo Hokkaido Forest. (Source: The University of Tokyo Hokkaido Forest)

During the 5th planning period, no new afforestation efforts were undertaken; rather, the management strategy concentrated solely on selective cutting of old-growth forests, utilizing natural regeneration (Tokyo University Forest in Hokkaido 1973). A working group system (refer to Chap. 8) was introduced, categorizing natural forests as the Selective Cutting Working Group (SCWG) and areas affected by forest fires as the Plantation Working Group. The SCWG was subject to a 25-year rotation period with an average cutting rate of 16%. The targeted annual tree harvest volume of 70,000 m^3, along with an additional 39,000 m^3 from land designated for cultivation, were nearly achieved as planned (The University of Tokyo Forests 2022b).

In 1931, the harvesting policy was realigned to focus on clearing old and over-mature stands. Consequently, the 5th management plan was partially revised; the rotation period was reduced to 20 years and the average cutting rate was set at 20%, including dead trees. It was believed that shortening the rotation period and increasing the cutting rate would still prevent overcutting while promoting the growth of remaining trees and maintaining the balance of the forest ecosystem (Tokyo University Forest in Hokkaido 1938). Despite these changes, the annual tree harvest volume declined to 56,000 m^3 in the latter half of the period due to reduced clear-cutting in areas designated for cultivation. Moreover, during the 5th planning period, extensive development efforts enhanced the utilization infrastructure, including the extension of the forest tramway to 50 km. Additionally, a sawmill laboratory with an annual capacity of 8500 m^3, an ironworks, and a power station (Fig. 2.7) were established (Tokyo University Forest in Hokkaido 1973; The University of Tokyo Forests 2022b).

Fig. 2.7 Spillway and generator of the power plant, 1938. (Photo by the UTHF, licensed under CC BY-NC-ND 4.0)

Under **the 6th management plan (1938–1947)**, natural forests were classified as the first working group, fire-regenerated forests as the second working group, and plantation forests as the third working group. The first working group was further divided into three categories: area for direct production (6000 ha), area for clearing decayed and dead trees (5000 ha positioned in the backcountry), and area for enhancing tree growth through silvicultural operations (9000 ha situated in the countryside). The average cutting rate was 25% for old-growth forests and 20% for forests that had already been selectively logged. The annual tree harvest volume was planned as 24,700 m^3, which corresponded to one-third of the actual harvest volume during the 5th planning period (The University of Tokyo Forests 2022b).

2.4 Forest Management During and After the War (1941–1957)

With the outbreak of the Pacific War in 1941, it became necessary to mobilize all domestic resources, both human and material, for the war effort (Tokyo University Forest in Hokkaido 1973; Kamata 2020). Large numbers of personnel were drafted, leading to severe labor shortages, and essential work supplies such as clothing, shoes, food, and horse feed were extremely scarce. To meet the quotas for supplying military materials, high-quality timber like monarch birch (*Betula maximowicziana*), Manchurian ash (*Fraxinus mandshurica*), and Yezo spruce (*Picea jezoensis*) was harvested intensively in the countryside forest. Additionally, 500 ha of clear-cutting were conducted in the Tairazawa and Kumanosawa areas for emergency cultivation from the wartime to postwar period. In the area for enhancing tree growth through silvicultural operations, firewood collection by colonists significantly reduced the growing stock, turning these areas into "forest for firewood and charcoal" (Tokyo University Forest in Hokkaido 1973).

The 6th management plan, which called for a drastic reduction in tree harvesting, did not align with the actual timber supply and demand at the time. Due to the need for high-quality timber for military use, clear-cutting for emergency cultivation, and concentrated logging in areas for enhancing tree growth through silvicultural operations, the management plan was disregarded (Tokyo University Forest in Hokkaido 1973). As a result, the average annual tree harvest volume increased to 38,200 m^3, representing a 55% increase over the original plan, and the countryside forest was extremely overcut (The University of Tokyo Forests 2022b).

After the war ended, forest management and experimental research were paused until 1948, when Kentaro Nakamura, Director of the Tokyo University Forests, established three guiding principles for the University Forests: model management, expansion of experimental research, and publication of results (Tokyo University Forest in Hokkaido 1973). **The 7th management plan (1948–1957)** was primarily organized to address the aftermath of irregular operations during the war (Shibata 1988). The forest area was divided into four working groups: The countryside forests (5000 ha), where accumulation had significantly decreased, were designated as the first working group, aiming to restore forest vigor through silvicultural operations, with a rotation period of 10 years and an average selective cutting rate of 12%. The backcountry forests (10,000 ha) were assigned to the second working group, with a rotation period of 20 years and a selective cutting rate of 25%, consistent with the previous planning period. The fire-regenerated forests (1900 ha) were designated as the third working group, with an average selective cutting rate 30% and a consolidation period of 20 years, primarily aimed at eliminating defective broadleaf trees. The plantation forests were categorized as the fourth working group, with only thinning conducted. The annual tree harvest volume was set at 39,200 m^3. However, in 1952, 764 hectares of forest land were released for emergency cultivation in the Kawamatsuzawa area, and 340,000 m^3 of wind-damaged and insect-damaged trees were caused by Typhoon Toyamaru in 1954. Consequently, the management plan was not implemented, and the annual tree harvest volume reached 80,200 m^3, twice the planned amount (Tokyo University Forest in Hokkaido 1999).

After the war, local demands for farmland liberation and guidance from related government agencies led to the sequential liberation of farmland and the creation of self-employed farmers between 1950 and 1964 (The University of Tokyo Forests 2022b). And 5318 ha of farmland were allocated to local farmers and 1306 ha of emergency farmland were transferred to repatriates after the war and to second- and third-generation local farmers. The land allocation also included 231 ha of road land and 73 ha of riverbed land (Tokyo University Forest in Hokkaido 2007). In 1952, Nobukiyo Takahashi, the fifth Director of the UTHF, formulated management reform policies. The main points of the policies were as follows: (1) to reduce logging operations and focus on cultivated forestry, (2) to adopt a flexible management plan, (3) to expand experiments and research, and (4) to open the door to the UTHF for contributing to the revitalization of the local forestry industry. Based on the policies, the forest tramway, ironworks, lumber laboratory, and power plant were successively abolished (The University of Tokyo Forests 2022b).

2.5 Beginning of "Stand-Based Silvicultural Management System" (1958–1980)

In 1958, the UTHF initiated **the 8th management plan (1958–1967)**, which aimed to achieve sustainable forest management based on the "Stand-Based Silvicultural Management System" (SSMS) (refer to Chap. 7 for an introduction). During the planning process, Director Takahashi determined that the best course of action for Hokkaido, where afforestation efforts are challenging, was to intensively manage natural forests using the selection system. This approach contrasted sharply with the silvicultural systems employed by the National Forest, which at the time advocated for the clear-cutting of large areas of natural forests to boost forest productivity (Shibata 1988).

This plan divided managed forests into three working groups. The first working group, comprising 9000 ha of forests situated in the countryside, implemented

Fig. 2.8 Selection cutting stand. Left: before cutting; Right: after cutting, Compartment 102. (Photo taken in 1970 by the UTHF, licensed under CC BY-NC-ND 4.0)

selection cutting[3] every 8 years with a cutting rate of 15% (Fig. 2.8). The supplemental planting stands and clear-cutting stands followed consolidation periods of 24 years (3 times the rotation period) and 32 years (4 times), respectively. The second working group, covering 11,000 ha of forests positioned in the backcountry, applied selection cutting at a 25% rate with a rotation period of 20 years. In the third working group, planted forests underwent thinning at a standard rate of 15%. During the 8th planning period, the planned annual harvest volume was 70,000 m^3 with an afforestation area of 91 ha. However, the results were 69,300 m^3 and 75 ha, respectively, achieving nearly the targeted harvest volume but only 87% of the planned afforestation area. During this period, in response to a declining labor force due to rapid economic growth, efforts were focused on establishing forest roads and introducing mechanized operations (The University of Tokyo Forests 2022b).

The 9th management plan (1968–1977, extended until 1980), developed in accordance with the SSMS, aimed to expand forest roads and consolidate operational zones to enhance efficiency and address challenges such as labor shortages (Shibata 1988). It sought to maintain the rotation periods and cutting rates established in the 8th management plan, with targets of an annual harvest volume of 74,000 m^3 and an afforestation area of 80 ha. However, actual results revealed that the annual harvest volume and afforestation covered only 68,000 m^3 and 59 ha, respectively, falling short of these goals. The deficits in harvest volume and afforestation areas were primarily attributed to a reduced labor force and significant wage increases (Shibata 1988). In response, the 9th planning period saw the introduction of bulldozers and other heavy machinery, aimed at boosting labor productivity and implementing measures to improve mobility. Although initially scheduled to conclude in 1977, the plan was extended to 1980, allowing time to develop strategies that addressed changing socioeconomic conditions during this transitional period. The establishment of 40 km of forest roads almost met the projections, achieving a road density of 29.6 m per hectare by 1980 (The University of Tokyo Forests 2022b). In 1972, the area on the southwest slope of Mt. Dairoku (1459 m) and Mt. Maeyama,

[3] The term "selection cutting" was used after the 8th management plan. Unlike selective cutting, which removes only large trees, selection cutting aims to stabilize forest structure and optimize volume production over repeated cutting cycles (Nyland 2016).

extending from the right bank of Honzawa (about 500 m above sea level) to the peak of Mr. Dairoku and covering approximately 1370 ha, was designated as a forest reserve (The University Forest in Hokkaido 1977).

2.6 Large Windthrow by a Typhoon and Management of Wind-Damaged Sites (1981–1985)

On August 23, 1981, the Typhoon No. 15 caused extensive wind damage, affecting 10,649 ha, nearly half of the total forested area (Fig. 2.9). A staggering 755,430 m^3 of wind-damaged trees were removed, representing approximately 18% of the total growing stock (Watanabe et al. 1990, refer to Chap. 13). The 10th management and experiment plan, initiated in the same year, was abruptly suspended and replaced by **a five-year interim management and experiment plan (1981–1985)** focusing on the treatment of windfall trees (The University of Tokyo Forests 2022b). The management strategy following the typhoon aimed at systematically clearing the damaged trees, considering their impact on future operations, vulnerability to insect infestations and wildfires, and the gradual deterioration of timber quality over time. For forest rehabilitation efforts, a targeted area of 3100 ha, severely affected by the typhoon, was earmarked for various interventions such as nursery cultivation of young trees, assisted afforestation, promotion of natural seeding through ground surface raking, and renewal of fallen trees, depending on their specific locations (The University of Tokyo Forests 2022b). The planned activities included the removal of 571,400 m^3 of windfall trees, establishment of 861 ha of afforestation, soil scarification of 150 ha of land utilizing heavy machinery, and construction of 10.3 km of forest roads annually, among other tasks. However, the results revealed that the actual volume of windfall trees removed exceeded expectations, reaching 810,000 m^3. Similarly, 273 ha of land were scarified, and 15.5 km of forest roads were constructed during the year, surpassing the planned targets. Nevertheless, the

Fig. 2.9 Uprooting of Yezo spruce in a mixed fir-spruce-birch forest damaged by the Typhoon No. 15 in 1981, Compartment 13. (Photo by the UTHF, September 1982, licensed under CC BY-NC-ND 4.0)

area of afforestation fell short, covering only 483 ha, indicating a deviation from the initial plan in this aspect (Shibata 1988; The University of Tokyo Forests 2022b).

2.7 Development of the Stand-Based Silvicultural Management System (1986–2010)

During the 10th through 12th planning periods, further developments were made based on the results of the SSMS over the previous 28 years, initiated by the 8th management plan (Tokyo University Forest in Hokkaido 1997). After completing the wind damage treatment in 1986, **the 10th management and experiment plan (1986–1995)** was implemented. In the 10th plan, the existing silvicultural management system was reorganized due to labor shortages and budget reductions. First, the third working group, composed of scattered plantations, was merged into either the first or second working group, which surrounded each plantation. For selection cutting in the first working group, the rotation period was extended from 8 to 10 years, and the average cutting rate increased from 15% to 16%. Additionally, the tree-marking criteria were changed so that large-diameter trees, which had previously been maintained, were harvested systematically and sequentially. This shift moved from a conventional single-tree selection system based on stand growth to actively promoting natural regeneration through a group selection system (Tokyo University Forest in Hokkaido 1997). In 1987, to evaluate the market value of high-quality hardwoods, the UTHF began selling monarch birch, Japanese oak (*Quercus crispula*), and Castor aralia (*Kalopanax septemlobus*) at the fancy log market in Asahikawa (Tokyo University Forest in Hokkaido 1997).

Furthermore, the consolidation period for supplemental planting stands was adjusted to align with the extended rotation period for selection cutting, shortening the previous 24-year period (8 years and 3 rotations) to 20 years (10 years and 2 rotations). The cutting rate was increased from 35% to 40% to enhance community-based silvicultural treatment, and useful hardwoods, which would be regenerated after supplemental planting, were also nurtured. Natural regeneration was further promoted using heavy machinery like a bulldozer equipped with a rake. The second working group also saw changes to promote natural regeneration, with the rotation period remaining at 20 years but the cutting rate reduced from 25% to 17%. For supplemental planting and clear-cutting stands, which had not previously been included in silvicultural operations, stand improvement efforts began in the 10th period with a consolidation period of 40 years. Additionally, the conventional classification method based on three stand types (selection cutting, supplemental planting, and clear-cutting) was improved by introducing more detailed forest types (Fig. 2.10). In the 10th period, the plan included harvesting 47,000 m^3 of trees, afforesting 113 ha, and establishing 20 km of forest roads (Tokyo University Forest in Hokkaido 1997). Consequently, the density of forest roads was increased to 39.2 m/ha. However, the actual harvest volume and afforestation fell short of the

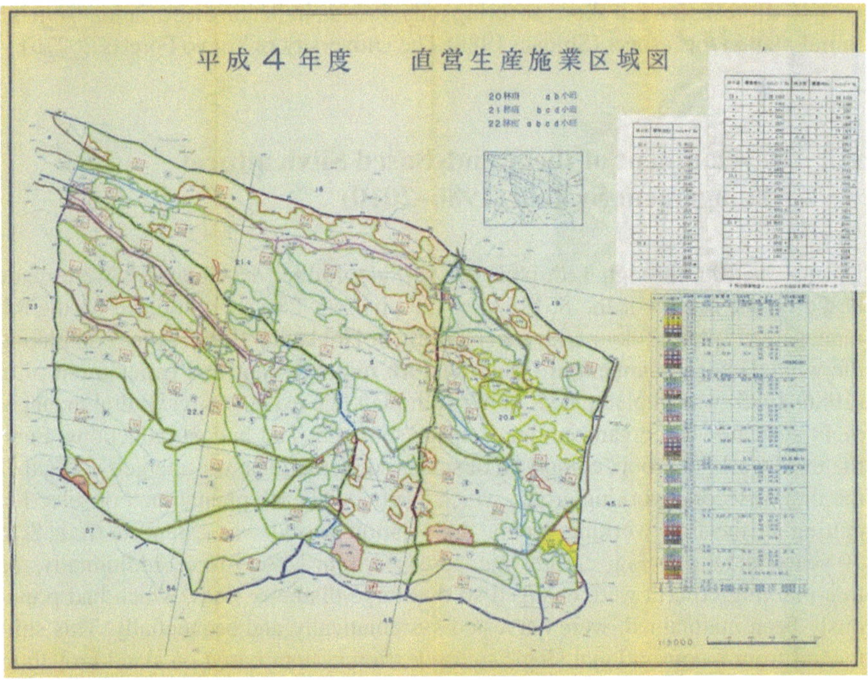

Fig. 2.10 Map of forest type classification for Compartments 20, 21 and 22, 1992. (Source: The University of Tokyo Hokkaido Forest)

targets, with the afforestation area achieving only 68% of the planned extent. Specifically, 353 ha were designated for supplemental planting and clear-cutting stands, but none were planted due to a lack of labor (The University of Tokyo Forests 2022b).

In **the 11th management and experiment plan (1996–2005)**, which mirrored the 10th plan, the rotation periods remained consistent with 10 years for the first working group and 20 years for the second working group. A new working group was introduced for secondary forests after fire, featuring a shorter 7-year rotation period and a 25% harvest rate, aimed at cultivating high-quality, large-diameter monarch birch timber. The planned annual harvest volume was set at 39,507 m^3, with 41 ha designated for afforestation. However, the actual figures were 36,300 m^3 of harvest and 32 ha of planting. Regarding infrastructure, 136 km of forest roads were planned, but only 39 km were successfully established, bringing the forest road density to 41.0 m/ha. No forest roads have been constructed since 2004 (The University of Tokyo Forests 2022b). In 1996, a grapple skidder was introduced into its logging operations and began the operational experiments with it (Tokyo University Forest in Hokkaido 1999). To enhance the accuracy of forest management information and systematize its advanced use, aerial photography, global positioning system (GPS) surveying, and geographic information system (GIS) were tested for use in silvicultural management operations (Tokyo University Forest in Hokkaido 2007).

Due to an excessive workload relative to available personnel at the time and the inadequate growth of small- and medium-diameter trees, which failed to regenerate sufficiently because of damage from mechanical logging with short rotation periods, significant changes were implemented in **the 12th management and experiment plan (2006–2010)**. These changes included extending the rotation period for the first working group to 15 years with the cutting rate for conifer-dominated selection cutting stands adjusted to between 10% and 16%. The rotation period for the second working group was maintained at 20 years, with a cutting rate of 10–17%. Although the plan originally projected an annual harvest volume of 26,635 m^3 and 34 ha of planting for the first five years, the actual outcomes averaged 23,100 m^3 of harvest and 18 ha of planting per year (The University of Tokyo Forests 2022b). Additionally, focusing on the properties of Japanese oak in Hokkaido as suitable for whisky barrel timber, an oak working group was designated for the 12th period (Tokyo University Forest in Hokkaido 2007). The full implementation of GIS and laser compass ground surveying systems for forest type classification mapping and surveying occurred in 2006 and 2008, respectively (Owari 2022). The 12th plan, initially scheduled to continue until 2015, was terminated prematurely in 2010 to synchronize the planning periods across the seven regional forests of the University of Tokyo Forests.

2.8 Digitalization of Natural Forest Management and Planning (2011–Present)

In **the 13th education and research plan (2011–2020)**, while maintaining the basic principles of the SSMS, the aim was to further develop and universalize the theory and methods of sustainable and adaptive forest management by building upon the improvements made in the 12th plan (The University of Tokyo Forests 2022b). The general experimental forests (forests subject to the SSMS application) were mainly divided into two working groups according to differences in site conditions and accessibility: the first Natural Forest Working Group (NFWG) situated in the countryside at low to medium elevations (11,200 ha) and the second NFWG positioned in the backcountry at medium to high elevations (7700 ha), and the rotation periods were set to 15 and 20 years, respectively, as in the 12th plan. In addition, separate working groups were established for secondary forests after fire, plantations (1100 ha), and broadleaf tree individuals in the Rouseppu area (Compartments 71–76) that had been severely disturbed by past fire events.

Each year, in silvicultural management, the managed forests were divided into seven forest types, each further subdivided based on dominant tree species, disturbance and regeneration history, and management methods. Definitions of each forest type and criteria/indicators for forest type classification were provided in the plan (The University of Tokyo Hokkaido Forest 2012). The cutting rates for conifer-dominated selection cutting forests were set at 16% in the first NFWG and 17% in the second NFWG. However, in forests where the growing stock of medium- and large-diameter trees was less than 250 m^3/ha, the cutting rate was reduced to 13% to enhance the forest resource. In conifer-dominated selection cutting forests with poor regeneration, the cutting rate was set at 10%. For broadleaf-dominated selection cutting forests, no harvesting was conducted this period in areas where it was deemed advantageous to increase volume and value growth, while a 10% cutting rate was set for forests requiring harvesting. The plan estimated an annual harvest of 24,184 m^3 and the planting of 26 ha. The actual average annual results were 24,136 m^3 of harvested timber and 23 ha of planting (The University of Tokyo Forests 2022b). In 2016, a typhoon accompanied by torrential rain caused a severe disaster, resulting in numerous forest road collapses (refer to Chap. 9).

Since the 2010s, the digitalization of natural forest management operations has advanced significantly. In 2012, global navigation satellite system (GNSS) surveying was introduced for classifying forest types, marking a significant technological integration (Owari 2022). Concurrently, the use of airborne laser scanning (LiDAR) began, and by 2019, LiDAR data ortho aerial photography covered the entire forest area (Owari 2020). In 2014, two types of unmanned aerial vehicles (UAVs), rotary-wing and fixed-wing, were introduced. Since then, the fleet, especially of rotary-wing UAVs, along with the variety of UAV models and units, has expanded (Fig. 2.11). By 2016, the forest management operations also saw the integration of electronic field books on table devices (Owari 2022).

The current 14th education and research plan (2021–2030) aims to contribute to natural forest management in Japan and globally (The University of Tokyo Forests 2022b). This plan follows the structure of the 13th plan, maintaining rotation periods of 15 years for the first NFWG and 20 years for the second NFWG (Fig. 2.12). The annual harvest volume is planned at 24,254 m^3, with 22.7 ha designated for afforestation. The average growing stock in natural forests managed under selection system has been increasing since the 10th period, reaching 306 m^3/ha at the beginning of the 14th period (The University of Tokyo Forests 2022b). The increase was particularly significant in plantations, where the average growing stock reached 206 m^3/ha at the beginning of the 14th period (The University of Tokyo Forests 2022a). The 14th plan aims to introduce innovative remote sensing technology for forest type classification surveys and plot inventory measurements,

Fig. 2.11 Acquisition of aerial images using an unmanned aerial vehicle (UAV). (Photo by the UTHF, licensed under CC BY-NC-ND 4.0)

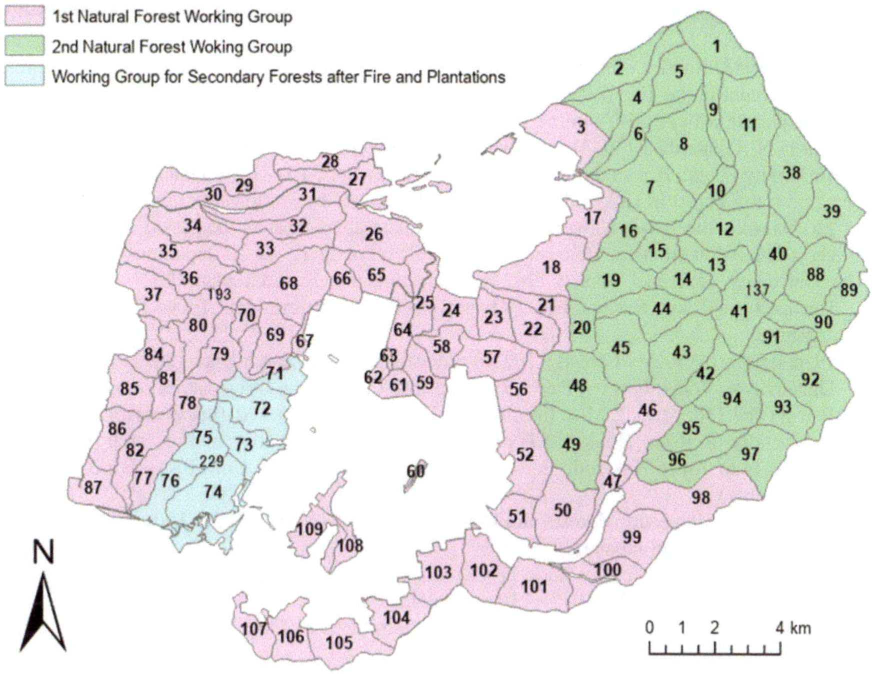

Fig. 2.12 Map of compartments and working groups in the University of Tokyo Hokkaido Forest as of 2024. (Source: The University of Tokyo Hokkaido Forest)

in preparation for potential future staff reductions (refer to Chap. 8). Low-cost dual-frequency real-time kinematic (RTK)-equipped GNSS receivers have been utilized for forest type classification surveying since 2021 (Owari 2022).

Acknowledgments The authors would like to thank current and former staff members of the UTHF for their continuous and great efforts in managing the forest. The description in this chapter mainly depends on Kamata (2020) and The University of Tokyo Forests (2022b).

References

Kamata N (2020) Overview, history, current status and issues of the University of Tokyo Hokkaido Forest. North For 71:81–86. (in Japanese)

Miura T (1923) History of the Hokkaido Forest until the 11th year of the Taisho era. Tokyo Imperial University Forest in Hokkaido. 142 pp. (in Japanese)

Nyland RD (2016) Silviculture: concepts and applications, 3rd edn. 680 pp. Waveland Press

Oohashi K, Arinaga A (2004) Reports of forest management and field survey by Tsuneo MIURA, the first director of the Tokyo University Forest in Hokkaido: observations of forest and forestry in Hokkaido 100 years ago by a forester. Misc Inform Tokyo Univ For 43:213–258. (in Japanese)

Owari T (2020) Applicability of airborne LiDAR and UAV in the stand-based silvicultural management system. North For 71:106–109. (in Japanese)

Owari T (2022) DX of natural forest management under the selection system. Sanrin 1656:36–42. (in Japanese)
Shibata S (1988) Studies on the management of natural forests: an experiment of natural forest management in the Tokyo University Forest in Hokkaido. Bull Tokyo Univ For 80:269–397. (in Japanese with English summary)
Takahashi N, Takahashi T (1944) A study on the colony of forestry labourer in the Tokyo Imperial University Forest, Hokkaido. Bull Tokyo Imper Univ For 34:1–81. (in Japanese)
The University Forest in Hokkaido (1977) Meteorological observations at different elevations of Mt. Dairoku, the Tokyo University Forest in Hokkaido: records in the period 1972–1975 and short discussion on them. Misc Inform Univ Tokyo For 21:22–47. (in Japanese)
The University of Tokyo Forests (2022a) Education and research plan (2021–2030) of the University of Tokyo Forests, Part 1 Overall plan. Misc Inform Univ Tokyo For 64:5–30. (in Japanese)
The University of Tokyo Forests (2022b) Education and research plan (2021–2030) of the University of Tokyo Forests, Part 3 Regional forest plans, The University of Tokyo Hokkaido Forest (The 14th period). Misc Inform Univ Tokyo For 64:103–190. (in Japanese)
The University of Tokyo Hokkaido Forest (2012) The 13th education and research plan of the University of Tokyo Hokkaido Forest (2011–2020). Misc Inform Univ Tokyo For 51:67–176. (in Japanese)
Tokyo Imperial University Forest, Hokkaido (1934) The outline of the University Forest in Hokkaido. 46 pp. (in Japanese)
Tokyo University Forest in Hokkaido (1938) The 6th silviculture plan of the Tokyo University Forest in Hokkaido (1938–1947). 210 pp. (in Japanese)
Tokyo University Forest in Hokkaido (1973) The 75th history of the Tokyo University Forest in Hokkaido. 116 pp. (in Japanese)
Tokyo University Forest in Hokkaido (1997) The 11th management and experiment plan of the Tokyo University Forest in Hokkaido (1996–2005). Misc Inform Univ Tokyo For 35:1–124. (in Japanese)
Tokyo University Forest in Hokkaido (1999) The 100th anniversary publication of the Tokyo University Forest in Hokkaido. 126 pp. (in Japanese)
Tokyo University Forest in Hokkaido (2007) The 12th management and experiment plan of the Tokyo University Forest in Hokkaido (2006–2015). Misc Inform Univ Tokyo For 46:215–350. (in Japanese)
Tokyo University Forests (1950) The 50th history of the Tokyo University Forest in Hokkaido. 61 pp. (in Japanese)
Watanabe S, Shibata S, Kawahara S, Shibano S, Kurahashi A, Satoo Y, Anazawa C, Takada N, Takahashi Y (1990) A memoir on the actual situation of the forest wind-damaged by the typhoon No. 15 in 1981 in the Tokyo University Forest in Hokkaido. Misc Inform Univ Tokyo For 27:79–221. (in Japanese)

Part II
Forest Ecosystem

Chapter 3
Location, Topography, Geology, and Soil

Nobuaki Tanaka

Abstract This chapter summarizes available information on location, topography, geology, and soil in the University of Tokyo Hokkaido Forest (UTHF), central Hokkaido Island, Japan. The UTHF covers two river basins, Nunobe and Nishitappu Rivers, both of which are tributaries of the Upper Sorachi River. There is an extensive flat or rolling terrain in the eastern UTHF. However, the topography along rivers is dissected by river erosion throughout the UTHF. The geology of the UTHF can be categorized into two types: (1) Mesozoic rocks formed during the Jurassic and Cretaceous periods and (2) welded tuff or volcanic rocks erupted from the Tokachidake Volcano Group since Neogene period. The welded tuff covers an extensive area of the UTHF, forming rolling or gently sloping lava plateaus, except in the dissected valleys. Soil types within the UTHF vary spatially, depending on climatic, vegetative, topographical, and geological conditions. Based on a previous comprehensive effort for mapping the forest soil in the UTHF, distinctive soil types distributing from the lowland to the upland of the UTHF are introduced in this chapter.

Keywords Hokkaido Island · Upper Sorachi River · Welded tuff · Mesozoic rocks · Soil type mapping

3.1 Location and Topography

The University of Tokyo Hokkaido Forest (UTHF) is located at Furano City, in the center of Hokkaido Island, Japan (43°10′–21′ N, 142°23′–41′ E; Fig. 3.1a). It lies to the southwest of the Tokachidake Volcano Group (Fig. 3.1a), extending about 20 km

N. Tanaka (✉)
The University of Tokyo Hokkaido Forest, Graduate School of Agricultural and Life Sciences, The University of Tokyo, Furano, Japan
e-mail: tnk-nobu@g.ecc.u-tokyo.ac.jp

T. Owari et al. (eds.), *Integrative Forest Management and Silviculture*,
Ecological Research Monographs, https://doi.org/10.1007/978-981-95-1017-7_3

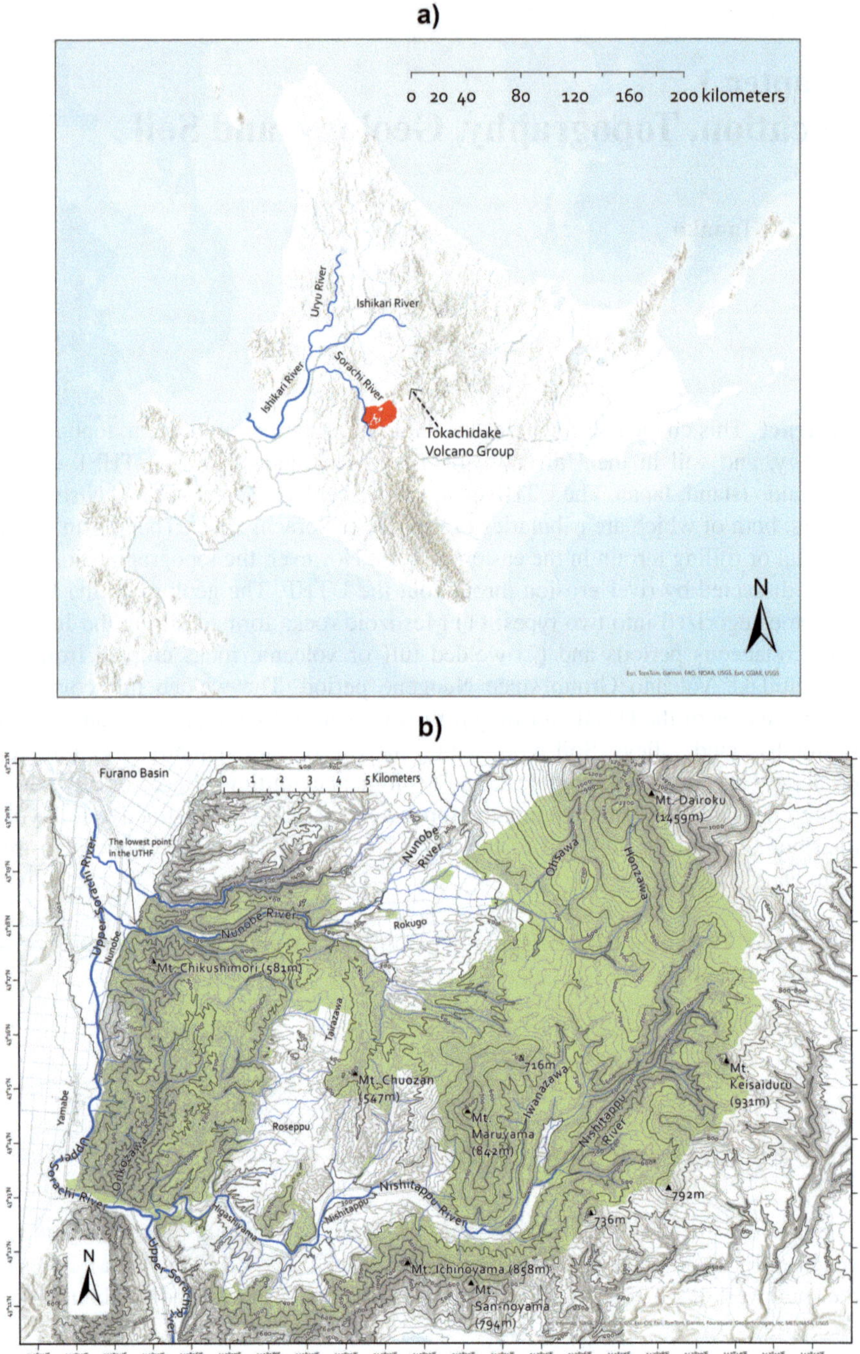

Fig. 3.1 Location (**a**) and topography (**b**) of the UTHF. The UTHF location is indicated by red and lime green areas, respectively, in (**a**) and (**b**). Basemap sources: Esri, Maxar, Airbus DS, USGS, NGA, NASA, CGIAR, N Robinson, NCEAS, NLS, OS, NMA, Geodatastyrelsen, Rijkswaterstaat, GSA, Geoland, FEMA, Intermap, and the GIS user community; Esri, TomTom, Garmin, FAO, NOAA, USGS, © OpenStreetMap contributors, and the GIS User Community

Fig. 3.2 Upper Nishitappu (**a**) and Nunobe (**b**) Rivers running in the UTHF (Photos by Nobuaki Tanaka). Middle reach of Nishitappu River (**c**, photo by the UTHF, May 22, 2008, licensed under CC BY-NC-ND 4.0) flowing along the UTHF boundary with a private cropland. Both the Nishitappu and Nunobe Rivers flow into the Upper Sorachi River. (**d**, photo by Nobuaki Tanaka, April 11, 2022)

east to west and 15 km north to south (Fig. 3.1b). All the UTHF areas are headwaters of the Sorachi River within the Ishikari River system (Fig. 3.1a). The UTHF encompasses two main river basins, Nunobe and Nishitappu, both of which flow into the Upper Sorachi River (Figs. 3.1b and 3.2d). Elevation of the UTHF ranges from the lowest point in the Nunobe area at 190 m a.s.l. to the highest point on Mt. Dairoku at 1459 m a.s.l. (Fig. 3.1b).

Historically, flat lowlands in and around the UTHF (Rokugo, Tairazawa, Roseppu, and Nishitappu; uncolored areas surrounded by the UTHF in Fig. 3.1b) were converted into farmlands after the settlement of people from mainland Japan or after World War II. As a result, the topography of the remaining UTHF exhibits a relatively complex terrain, compared to the farmlands nearby (Fig. 3.1b). However, there is an extensive flat or rolling terrain in the eastern UTHF (Figs. 3.1b and 3.3a), though, throughout the UTHF, the topography along rivers is commonly dissected by river erosion (Figs. 3.1b and 3.3b). A flat lowland remains in the UTHF exclusively at Tairazawa (Figs. 3.1b and 3.3c), which is currently managed as a forest reserve by the UTHF (Chap. 5).

Fig. 3.3 (**a**) The highest mountain in the UTHF, Mt. Dairoku (photo by the UTHF (licensed under CC BY-NC-ND 4.0, refer to also Fig. 3.1b). In the foreground of its summit, typical topography in the eastern UTHF, such as flat and rolling terrains, can be seen. (**b**) A steep slope topography besides Nunobe River (photo by Nobuaki Tanaka, December 27, 2024). (**c**) The remaining flat lowland topography at the basin bottom covered by a wetland forest. (Photo by Nobuaki Tanaka, July 1, 2022)

3.2 Geology

The geology of the UTHF can be categorized into two types (Fig. 3.4): Mesozoic rocks formed under the sea, specifically during the Jurassic and Cretaceous periods (approximately 150 to 65 million years ago), and pyroclastic flow deposits or volcanic rocks erupted from the Tokachidake Volcano Group since Neogene period (2.7 million years ago) to the present. The Mesozoic rocks are divided into three groups: the Hidaka Group, part of the Hidaka Belt; the Sorachi Group; and the Yezo Group,

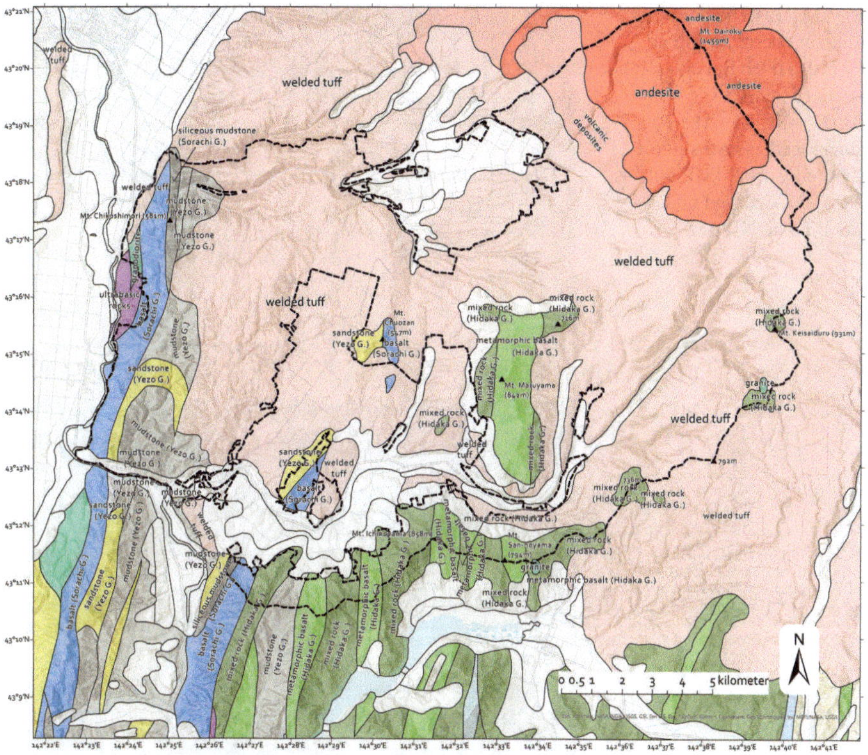

Fig. 3.4 Geological map of the UTHF. The UTHF boundary is delimited by black dash line. Deposits/rocks derived from Tokachidake Volcano Group and the Mesozoic rocks are colored, respectively, in red-to-pink gradients and others. For the Mesozoic rocks, the formation group is given in parenthesis. Note that quaternary deposits are not illustrated on the map (white-colored areas). The map is produced based on a vector style geological data, which is provided by Geological Survey of Japan, AIST (2023). Basemap sources: Esri, Maxar, Airbus DS, USGS, NGA, NASA, CGIAR, N Robinson, NCEAS, NLS, OS, NMA, Geodatastyrelsen, Rijkswaterstaat, GSA, Geoland, FEMA, Intermap, and the GIS user community; Esri, TomTom, Garmin, FAO, NOAA, USGS, © OpenStreetMap contributors, and the GIS User Community

both belonging to the Sorachi-Yezo Belt (Takashima et al. 2002). Each group exhibits a belt-like arrangement extending north-south (Fig. 3.4). These Mesozoic rocks mainly appear on the surface in the western and southern margins of the UTHF, while the rest of the area is widely covered by Tokachidake welded tuff or Dairokusan lava (Fig. 3.4, Hashimoto 1953; Osanai et al. 1968; Takashima et al. 2002).

The Hidaka Group is an accretionary complex that formed from the early Cretaceous to the early Tertiary period. It is composed of metamorphosed sandstones, mudstones, cherts, greenstones, and limestones (Hashimoto 1953; Osanai et al. 1968; Takashima et al. 2002). Dolomitized limestones are exposed in southern Nishitappu and are currently managed as a dolomite mine (Fig. 3.5a). The Sorachi Group consists of formations derived from submarine volcanic activity that occurred

Fig. 3.5 (**a**) The dolomite mine in operation (photo by Prof Hideo Sakai, licensed under CC BY-NC-ND 4.0). (**b**) The abandoned asbestos mine. (Photo by the UTHF, Jul. 28, 2006, licensed under CC BY-NC-ND 4.0)

from the late Jurassic to the early Cretaceous. Serpentinite intrusions, found at the base of the Sorachi Group in the UTHF were mined for asbestos until 1969 (Fig. 3.5b). The Yezo Group comprises sedimentary deposits from a forearc basin and predominantly consists of sandstones and mudstones (Takashima et al. 2002).

The pyroclastic flow deposits or volcanic rocks from the Tokachidake Volcano Group can be classified into two categories: the lower Tokachidake welded tuff and the upper Dairokusan lava. The Tokachidake welded tuff consists of ignimbrites and dacites that erupted from the end of the Pliocene to the early Pleistocene (Hashimoto 1953; Osanai et al. 1968). This welded tuff covers an extensive area of the UTHF, forming rolling or gently sloping lava plateaus, except in the dissected valleys (Figs. 3.1b and 3.3b). The Dairokusan lava is found only in the northeastern part of the UTHF and is primarily composed of andesite (Hashimoto 1953; Osanai et al. 1968).

3.3 Soil

Soil types within the UTHF vary based on climatic, vegetative, topographical, and geological conditions. The most comprehensive investigation of forest soil in the UTHF was conducted by Asahi (1963), who classified the soil into seven types: mountain black soil, podzol, dark forest soil, brown forest soil, swampy black forest soil, glei soil, and other minor soils.

Mountain black soil is characterized by a thick, black-colored humus layer and is typically found on ridges or mountain slopes at altitudes above 800 m a.s.l. (purple-colored areas in Fig. 3.6). Podzol can be distributed in the altitudinal range between 500 and 700 m a.s.l., but the distribution is limited mostly to the upper Nishitappu River in the UTHF (red-colored area in Fig. 3.6). Asahi (1963) reported that the podzol distribution was influenced by microtopography and can typically be found on gentle ridges or in topographical inflection areas between gentle and steep slopes (i.e., convex), and confirmed typical podzolization processes in the soil, that is, the leaching and accumulation of minerals and organic matters. Vegetation on podzol soils is predominantly characterized by *Abies sachalinensis* and *Piecea jezoensis*, almost always accompanied by *Picea glehnii* (Asahi 1963).

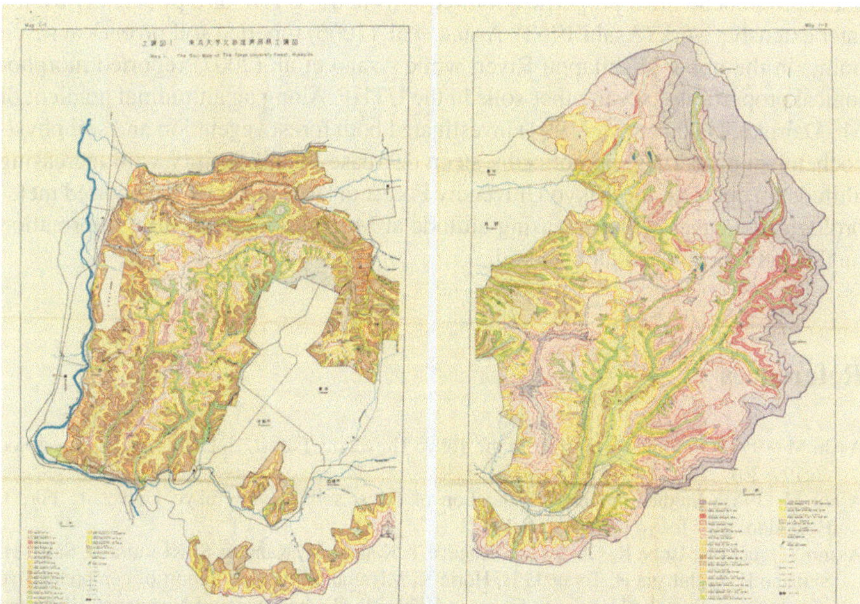

Fig. 3.6 The extensive soil type mapping work by Asahi (1963): This is reproduced with the publishers' permission

Dark forest soil and brown forest soil are the predominant soil types in the UTHF, covering elevations between 500 and 800 m a.s.l. Dark forest soil (pink-colored areas in Fig. 3.6) generally occurs at intermediate altitudes between mountain black soil on upper ridges and brown forest soil (yellow- or brown-colored areas in Fig. 3.6) on lower slopes. While the dark forest soil shows signs of weak podzolization such as a mineral mobility from A to B Horizons, the brown forest soil has no indication of podzolization (Asahi 1963). In addition, while almost all tree species in the UTHF are found in the brown forest soil, they are limited to conifers and Elman's birch, *Betula ermanii*, in the black forest soil (Asahi 1963).

Swampy black forest soil is distributed along riparian areas in the UTHF (green-colored areas in Fig. 3.6), typically accompanying various broadleaved tree species such as Manchurian ash, Japanese elm, Manchurian alder, katsura tree, painted maple, and goat willow. The glei soil appears exclusively in a flat lowland in the UTHF (Tairazawa, a gray-colored area in Fig. 3.6), where persistent high groundwater produces chemically reduced environment in the soil (Asahi 1963). Both the swampy black forest soil and the glei soil are relatively fertile soils compared to other soil types in the UTHF (Asahi 1963).

While not exhaustive, additional information on the soil in the UTHF can be found in various literature sources (Asahi 1954; Asano et al. 2006, 2007; Nakata et al. 1994; The Tokyo University Forest in Hokkaido 1977). Asahi (1954) investigated physicochemical properties of three soils in the UTHF as a pilot study for the later extensive work (Asahi 1963). Asano et al. (2006) reported soil profiles at seven points in the upper Nishitappu River, while Asano et al. (2007) reported morphological properties of seven other soils in the UTHF. Along an altitudinal gradient of Mt. Dairoku, Nakata et al. (1994) investigated both forest vegetation and soil physicochemical properties: he noticed a steep decrease in soil fertility with increasing altitude. In contrast, the Tokyo University Forest in Hokkaido (1977) reported meteorological changes with increasing altitude at Mt. Dairoku, with little information on the soil type at each plot studied.

References

Asahi M (1954) On some forest soils in the Tokyo University Forest, Hokkaido. J Jpn Forest Soc 36:199–201

Asahi M (1963) Studies on the classification of forest soils in the Tokyo University Forest, Hokkaido. Bull Tokyo Univ For 58:1–132

Asano Y, Iguchi K, Isobe R, Ohkawa A, Kagaya T, Kajiura M, Kamoda S, Ki-mura N, Sakai H, Shibano H, Tachibana H, Teragaki H, Horie K, Miyamoto Y (2006) Report of comprehensive watershed study at the University Forest in Hokkaido (1): headwater of Nishitappu River and Senninkyo. Miscellaneous Information of The University of Tokyo Forests 45:235–270

Asano Y, Inutsuka M, Ohkawa A, Kagaya T, Kajiura M, Kimura N, Saito N, Sa-kai H, Shibano H, Tachibana H, Nishida K, Horie K, Miyamoto Y, Waki S (2007) Comprehensive study on stream environments and ecosystems at forested mountainous watershed: report of comprehensive watershed study at the University Forest in Hokkaido. Miscellaneous Information of The University of Tokyo Forests 46:123–158

Geological Survey of Japan, AIST (2023) Seamless digital geological map of Japan V2 1:200,000, Original edition. https://gbank.gsj.jp/seamless/. Accessed 17 Jan 2024
Hashimoto W (1953) Explanatory text of the geological map of Japan, Yamabe (scale 1:50,000). Hokkaido Development Agency. 101 pp
Nakata M, Tanaka H, Yagi H (1994) Altitudinal changes in vegetation and soils on Mt. Dairoku, Central Hokkaido. Jpn J Ecol 44:33–47
Osanai H, Sako S, Matsui K, Matsushita K (1968) Explanatory text of the geo-logical map of Japan, Nishitappu (scale 1:50,000). Geological Survey Hokkaido. 28 pp
Takashima R, Miyamoto Y, Nichi H, Yoshida T (2002) Geology and stratigraphy of the Sorachi and Yezo Groups in the Tokyo University Forests in Hokkaido. Japan Bull Tokyo Univ For 108:57–76
The Tokyo University Forest in Hokkaido (1977) Meteorological observations at different elevations of Mt. Dairoku, the Tokyo University Forest in Hokkaido. Bull Tokyo Univ For 21:22–47

Chapter 4
Climate and Hydrology

Nobuaki Tanaka

Abstract There are distinct seasonal changes in both precipitation and air temperature in the UTHF. Annual mean air temperature has increased since 1990s, to an extent similar or slightly exceeding the reported global warming rate of 1.1 °C. Mean annual precipitation in the UTHF is lower than the nationwide average, due to low frequencies of both Typhoon arrival and Baiu front stagnation. Snow cover period appears at the end of November and lasts until the beginning of April, with a mean maximum snow depth of 85.6 cm typically observed in late February. Seasonal fluctuations in stream discharge are driven primarily by snowmelt in spring and rain-runoff responses in summer and autumn. Both low and flood flows are hydrological risks in the region. Catchment geology affects particularly on summer low flow. While spring flows may contribute to stream discharge, no comprehensive investigation has yet been conducted in the UTHF. Although fragmented, available information on streamwater chemistry in and around UTHF suggested a strong influence of catchment geology on the chemistry.

Keywords Warming · Snowpack · Snowmelt · Summer low flow · Geology

4.1 Climate

The climate prevailing in the UTHF is continental and subarctic, with a large seasonal variation in air temperature (Fig. 4.1). Annual mean air temperature at the arboretum (Yamabe, 226 m a.s.l.) is 6.6 °C for the period 2011–2020, with the maximum and the minimum instantaneous temperatures of 35.9 °C and − 25.7 °C, respectively. The long-term mean air temperature exhibits two fluctuating patterns (Fig. 4.2): (1) a relatively constant trend from 1926 to 1989, with an average

N. Tanaka (✉)
The University of Tokyo Hokkaido Forest, Graduate School of Agricultural and Life Sciences, The University of Tokyo, Furano, Japan
e-mail: tnk-nobu@g.ecc.u-tokyo.ac.jp

T. Owari et al. (eds.), *Integrative Forest Management and Silviculture*,
Ecological Research Monographs, https://doi.org/10.1007/978-981-95-1017-7_4

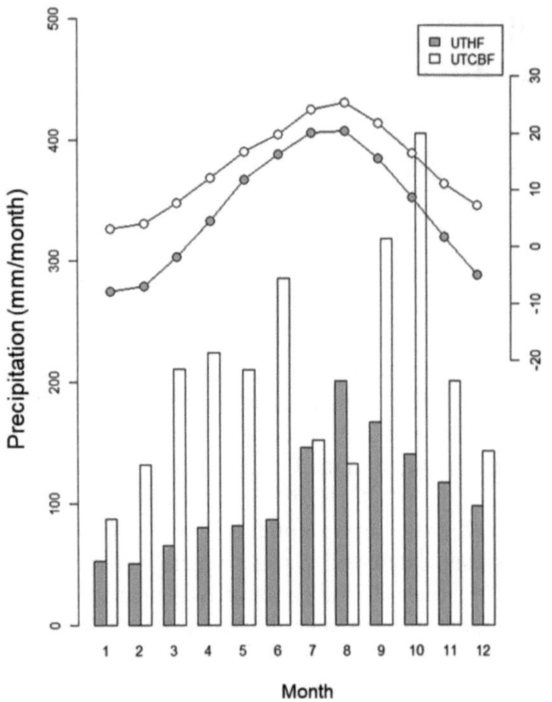

Fig. 4.1 Mean monthly temperature (solid circle) and precipitation (gray bar) at the UTHF. For comparison, the same variables but at the UTCBF (see text) are also shown by open circle and blank bar

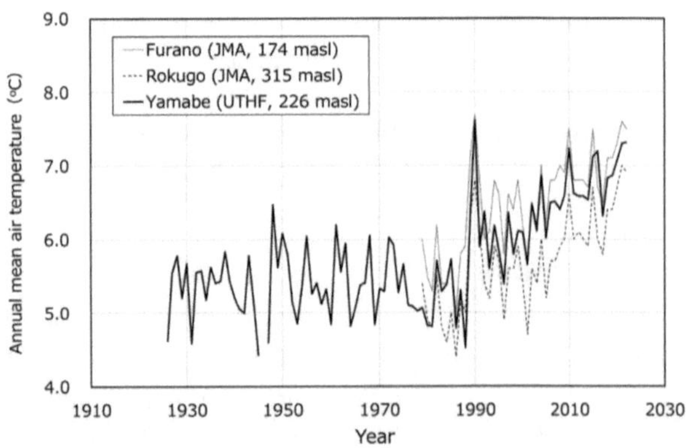

Fig. 4.2 Time series of annual mean air temperature at the UTHF (Yamabe). For comparison, the same variable but at nearby Japan Meteorological Agency (JMA) stations are also shown. (Furano and Rokugo)

temperature of 5.4 °C and (2) an upward trend starting from 1990. The recent mean temperature (6.6 °C) is 1.2 °C higher than the average during the period 1926–1989. For confirmation, the temperature increase at Yamabe is compared with that of surrounding areas (Fig. 4.2). Although the baseline air temperatures at Furano (174 m a.s.l.) and Rokugo (315 m a.s.l.), both of which were measured by Japan Meteorological Agency (JMA), differ from that of Yamabe due to altitudinal differences, a similar increasing trend to that observed at Yamabe is also evident at the two JMA stations (Fig. 4.2). The similarity makes the observed increase in air temperature at Yamabe during the recent three decades more convincing. For comparison, according to the most recent Intergovernmental Panel on Climate Change report (Gulev et al. 2021), the global surface temperature during the period 2011–2020 was 1.1 °C higher than preindustrial (1850–1990) levels. The recent air temperature increment at Yamabe appears to be compatible to or slightly greater than the global temperature increase (Fig. 4.2).

The mean annual precipitation at Yamabe is 1196 mm, which is much smaller than a nationwide average of approximately 1700 mm. The annual precipitation at Yamabe unevenly distributed over the year (Fig. 4.1), with a single peak in August (201 mm/month). For comparison, Fig. 4.1 includes rainfall seasonality at a station in The University of Tokyo Chiba Forest (UTCBF), which locates on Pacific Ocean side of Boso Peninsula, the main island of Japan. Annual total precipitation at the UTCBF station is 2501 mm and the seasonality shows a bimodal distribution (Fig. 4.1). The earlier precipitation peak in June and the later peak in October at UTCBF is caused, respectively, by stagnant weather front (Baiu front) and frequent arrivals of tropical cyclone (Typhoon). The UTHF is not directly affected by the two precipitation systems, and thus, lack of the two peaks may be the most notable features of the UTHF precipitation seasonality. Note that, though frequency is lower than the UTCBF, the UTHF can occasionally be prone to typhoon (see Chap. 13).

The snow cover period at Yamabe typically appears at the end of November and lasts until the beginning of April, during which almost all precipitation is brought by snowfall. Mean maximum snow depth is 85.6 cm at Yamabe, which is usually observed in the late February. Kameda et al. (2023) analyzed a long-term JMA record over each region of Hokkaido, and they found no significant increasing/decreasing trend in the maximum snow depth in surrounding regions around the UTHF.

Historically, there were other meteorological stations in the UTHF besides Yamabe. Daily meteorological measurements began at Nishitappu and Rokugo (Fig. 4.3) in 1919 and 1920, respectively, while measurements at Yamabe started in 1926. The daily measurements were maintained manually by the technical staff, and part of the data has been published (e.g., Anonymous 1940). However, the operation of the two stations by the UTHF terminated in 1975 and JMA succeeded the Rokugo station in later year (1978) with an automated measurement system. Since 1980s, the manual measurements at Yamabe had gradually been shifted to the automated measurement using electric data loggers and a fully automated system using a data logger (CR10X, Campbell Scientific, USA) eventually introduced to Yamabe station in 2000. In order to cover the geographical gap in the meteorological data, the UTHF added an upland meteorological station at Maeyama in 2004 (610 m a.s.l., Fig. 4.3). The meteorological data is currently available at the University of Tokyo Forests (UTF) website (UTF 2024a).

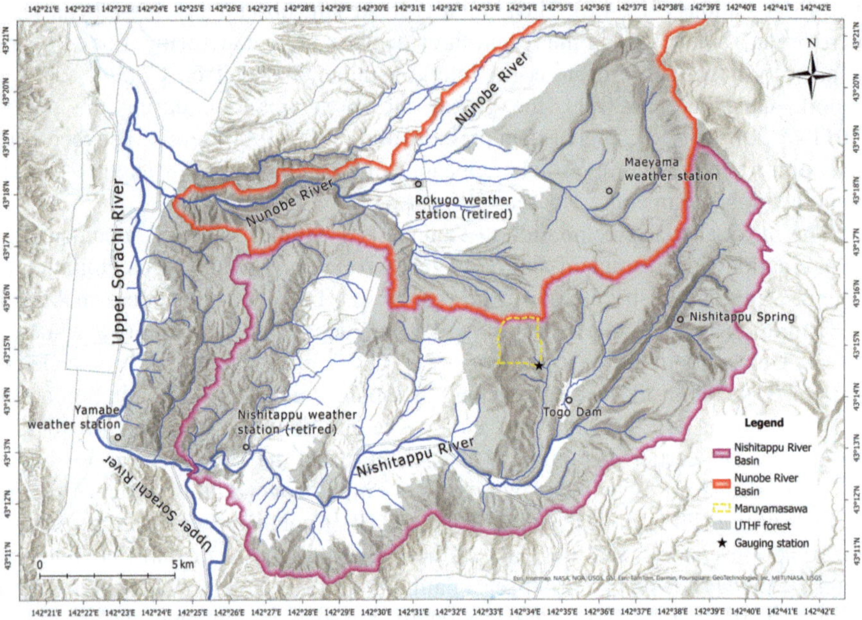

Fig. 4.3 Stream channels in and around the UTHF. The map is produced based on polygon data, which is provided by UTHF. Basemap sources: Esri, Maxar, Airbus DS, USGS, NGA, NASA, CGIAR, N Robinson, NCEAS, NLS, OS, NMA, Geodatastyrelsen, Rijkswaterstaat, GSA, Geoland, FEMA, Intermap, and the GIS user community; Esri, TomTom, Garmin, FAO, NOAA, USGS, © OpenStreetMap contributors, and the GIS User Community

4.2 Hydrology

Topographically, there is a gentle ridge running east-west throughout the UTHF (Chap. 3) and the ridge splits the UTHF into two river basins: Nishitappu and Nunobe River Basins (Fig. 4.3). Both rivers are major tributaries of upper Sorachi River, while many smaller rivers along the western margin of the UTHF confluent directly with the upper Sorachi River (Fig. 4.3). Headwaters of these major and smaller river basins are the UTHF forest (Fig. 4.3), while there are also privately owned farmlands, particularly in flat lowland areas of the basins.

As part of long-term ecohydrological monitoring of the UTHF, a streamflow gauging station in upper Nishitappu River Basin has been operated since 1983 (Maruyamasawa in Figs. 4.3 and 4.4). The station calculates flow rate from water level of a stilling pool, which is made by damming the streamflow by a concrete weir with a shape-crested iron notch (Shibano et al. 1988, Fig. 4.4). Precipitation has been measured at a nearby open site (Togo Dam in Fig. 4.3), which is closer than Yamabe. The measured data is available on UTF webpage (UTF 2024a). Fig. 4.5 shows a portion of the data, illustrating the seasonality in discharged streamflow from Maruyamasawa in 2022. Low flow in the mid-winter was followed by

Fig. 4.4 Bird-eye view of Maruyamasawa streamflow gauging station. (Photo by the UTHF, licensed under CC BY-NC-ND 4.0)

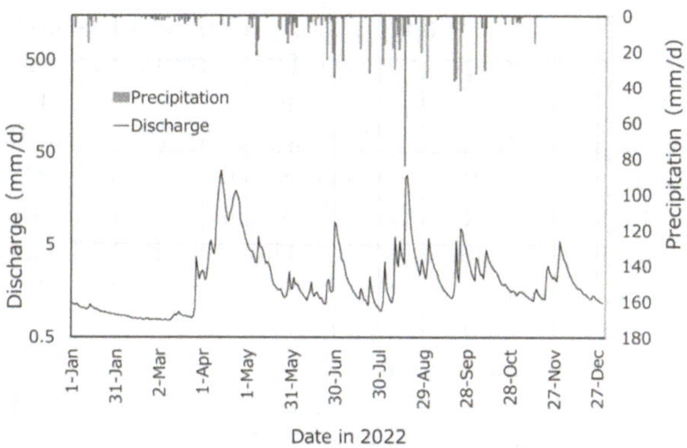

Fig. 4.5 Seasonal changes in stream discharge at Maruyamasawa and in precipitation at Togo Dam

about one-month long high flow in the absence of high precipitation (Fig. 4.5). The long-lasting high flow is produced by gradual progression of snowmelt in the watershed. Though not clear in the 2022 case, the snowmelt discharge is typically followed by summer low flow, which might be caused by low precipitation (Fig. 4.5), relatively high temperature (i.e., high evaporative demand of the atmosphere) and active water use by leaved trees in the watershed. In autumn, the streamflow discharge exhibits repeated peak flow in response to rainfall and recession flow after the peak flow. Note that the autumn period from late August to October is a typical season of flood flow risk. One of the most serious floods occurred in August in 1962 at Higashiyama village (Higashiyama Local History Editing Committee 1989), where more than 100 houses in the village were flooded by river blockage by driftwood at the bridge pier. Another serious flood flow occurred in late August in 2016, where four typhoons approached the UTHF in immediate succession, bringing more than 400 mm/month rainfall to the UTHF. The heavy rains caused at least 79 failures of slopes, forest roads, and bridges, within the UTHF (Isozaki et al., unpublished, refer to also Fig. 9.4 in Chap. 9). Additionally, levee breaches along the upper Sorachi River triggered a flooding disaster at a town next to the UTHF (Okuda et al. 2017).

Summer low flow can also be a hydrological risk to the local economy, because streamflow in the UTHF is intensively utilized as a water resource for agriculture

(Chap. 16). Based on a summer flow investigation conducted in the late July, 2005, Asano et al. (2009) suggested that the summer flow depended on catchment geology. More specifically, serpentinite and limestone catchments produced lowered flow rate (0.38 and 0.36 mm/day, respectively), while andesite and welded tuff catchments in the upper Nunobe River Basin showed much higher flow (2.42 and 2.37 mm/day, respectively). The difference in the summer flow may be explained by differential water retention capacities between geological types. Nakamura (1995), on the other hand, reported that another welded tuff catchment in the upper Nishitappu River Basin had much lower low flow (a 10-year mean of 0.35 mm/day). Further studies may be necessary to fully understand spatial-temporal variability of summer low flow in the UTHF.

Spring flow from an underground aquifer is an important hydrological process in the UTHF, contributing to the regulation of baseflow in the UTHF streams. The most prominent spring in the UTHF (Nishitappu Spring, Fig. 4.6) is located near the main channel of the upper Nishitappu River (Fig. 4.3), where apparently constant spring flow from cracks of welded tuff can be found. The flow rate was estimated at 0.41 m^3/s on average (Asano et al. 2011). Asano et al. (2011) estimated travel time of the groundwater at around 30 years and furthermore estimated the groundwater volume at 39 million m^3. In addition to the Nishitappu Spring, there are a lot of smaller springs in the UTHF (Asano et al. 2006; Iwasaki et al. 2023). All these springs contribute to mitigating summer low flow and to cooling of hot streamwater in summer (Iwasaki et al. 2023).

Precipitation and streamwater chemistry have been monitored monthly at Togo Dam and Maruyamasawa (Fig. 4.3), respectively, since 2003, as part of ecohydrological monitoring of the UTHF. Due to poor accessibility, the monitoring at Maruyamasawa is suspended during the snowy season, the chemistry data has been made available annually, alongside data from other UTF streams, on a website (UTF 2024b). According to UTF (2024b), no notable changes in the precipitation and streamwater chemistry have been observed over the past two decades. Sulfate concentration in streamwater is of a concern for humans downstream, because the UTHF streamwater is important freshwater for irrigation (Chap. 16) and some streams near the UTHF show extremely high sulfate concentration (up to 115 mg/L) due to ongoing volcanism around Tokachidake Volcano Group (Fig. 3.1a, Nishi

Fig. 4.6 Nishitappu Spring in the UTHF. (Photo by Nobuaki Tanaka)

et al. 1974; Shintani and Morino 2024). In this respect, UTF (2024b) reports much lower sulfate concentration for Maruyamasawa (typically less than 3 mg/L).

An extended short-term campaign investigating streamwater chemistry in the UTHF was conducted in the summer low flow season in 2005 (Asano et al. 2007). According to the results, bedrock appears to influence the streamwater chemistry. For example, a catchment partially underlain by limestone showed much higher calcium concentrations (about 20 mg/L) than others in the UTHF. Additionally, a serpentine catchment exhibited relatively high magnesium (up to 23.5 mg/L) and calcium (about 10 mg/L) concentrations compared to other UTHF catchments. The elevated magnesium concentration in serpentine catchments is attributed to bedrock weathering (Yamazaki et al. 2019) and has also been observed in other regions in Hokkaido (Yamazaki et al. 2019).

References

Anonymous (1940) *Enshurin Kishou Houkoku* [University Forest Meteorological Report], Miscellaneous Information, the University of Tokyo Forests, 2:1–240

Asano Y, Iguchi K, Isobe R, Ohkawa A, Kagaya T, Kajiura M, Kamoda S, Kimura N, Sakai H, Shibano H, Tachibana H, Teragaki H, Horie K, Miyamoto Y (2006) Report of comprehensive watershed study at the University Forest in Hokkaido (1): headwater of Nishitappu River and Senninkyo. Miscellaneous Information of The University of Tokyo Forests 45:235–270

Asano Y, Inutsuka M, Ohkawa A, Kagaya T, Kajiura M, Kimura N, Saito N, Sakai H, Shibano H, Tachibana H, Nishida K, Horie K, Miyamoto Y, Waki S (2007) Comprehensive study on stream environments and ecosystems at forested mountainous watershed: report of comprehensive watershed study at the University Forest in Hokkaido. Miscellaneous Information of The University of Tokyo Forests 46:123–158

Asano Y, Shibano H, Tanaka N, Ohkawa A, Asai K (2011) Mean transit time estimate of the spring at the headwater of Nishitappu River using chlorofluorocarbons (CFCs). Miscellaneous Information of The University of Tokyo Forests 50:41–48

Gulev SK et al (2021) Changing state of the climate system. In: Climate change 2021: the physical science basis. Contribution of working group I to the sixth assessment report of the intergovernmental panel on climate change [Masson-Delmotte, V., et al. (eds.)]. Cambridge University Press, Cambridge/New York, pp 287–422

Higashiyama Local History Editing Committee (ed) (1989) *Higashiyama Kyoudoshi* [Higashiyama Local History], Higashiyama 80th Anniversary Commemorative Project Executive Committee, Japan, 594 pp

Iwasaki K, Fukushima K, Nagasaka Y, Ishiyama N, Sakai M, Nagasaka A (2023) Real-time monitoring and postprocessing of thermal infrared video images for sampling and mapping groundwater discharge. Water Resour Res 59:e2022WR033630

Kameda T, Kuwasako T, Shirakawa T (2023) Characteristics of snow depth and period variations in snowy areas in Japan for 60 years from 1962 to 2021 winter season using annual average snow depth and annual maximum snow depth. Seppyou 85:199–222

Nakamura K (1995) Three Hokkaido maps and two flow duration tables. Monthly report of the Civil Engineering Research Institute 506:26–29

Nishi T, Sugimoto K, Takahashi M, Sakai T, Yonekawa M, Miyagacho S, Kato G (1974) On the water examinations of the rivers at the piedmont district of Mt. Tokachi in and around Kamifurano Cho. Research Bulletin of Obihiro University 8:319–343

Okuda J, Shimizu Y, Kyuka T, Iwasaki T, Ishida Y, Kuwamura S (2017) The study on 2016 August flood disaster and levee breach at the Sorachi River. Adv River Engineer 23:61–66

Shibano H, Mikami K, Nishio K (1988) Runoff analysis in the central Hokkaido considering the snow coverage and melting Bull Tokyo University Forests 80:129–155

Shintani T, Morino Y (2024) Geochemical characteristic and spatial distribution of groundwater and surface water in Frano area, Hokkaido. Research Institute of Energy, Environment and Geology Bulletin 3:25–44

UTF (2024a) *Kishou Kansoku Data* [Meteorological data], https://www.uf.a.u-tokyo.ac.jp/research_division/data/kishou/index.html. Last accessed in 20th Sept 2024

UTF (2024b) *Suimonkansoku Suishitsubunseki Data* [Hydrological and water quality data], https://www.uf.a.u-tokyo.ac.jp/research_division/data/suimon_suishitsu/index.html. Last accessed in 15th Oct 2024

Yamazaki A, Hatanaka T, Nakatani N, Horie K, Mizuno N, Hobara S (2019) Relationship between river water quality and soil characteristics in serpentinite area of Hokkaido. J Rakuno Gakuen Univ (Nat Sci) 44(1):19–27

Chapter 5
Forest Vegetation and Ecological Studies

Satoshi N. Suzuki and Susumu Goto

Abstract The University of Tokyo Hokkaido Forest (UTHF) is located in a hemiboreal mixed forest zone and is predominantly characterized by boreal conifers such as *Abies sachalinensis*, *Picea jezoensis*, and *Picea glehnii* and cool-temperate deciduous broadleaves such as *Tilia japonica*, *Acer pictum*, *Betula* spp., and *Quercus crispula*. As an ecotone, the forest composition within the UTHF varies with elevation, transitioning from a cool temperate deciduous broadleaf-dominant forest at lower elevations to an alpine dwarf pine (*Pinus pumila*) forest at the highest point. The vegetation associated with specific disturbances, topography, or geography serves as a habitat for diverse species. Over the span of more than 60 years, the UTHF has conducted long-term measurements of forest trees. In addition to numerous small forest plots (typically smaller than 1 hectare), there are two large-scale long-term ecological research (LTER) plots: Maeyama LTER (36.25 hectares) and Iwanazawa LTER (18.75 hectares). These extensive monitoring efforts have elucidated the dynamics and environmental changes in hemiboreal forests. The UTHF has established several experimentally designed study plots utilized for assessing biodiversity-ecosystem functional relationships and the genetic local adaptations of trees to elevational gradients. Long-term observations of plant phenology have also been conducted, providing robust evidence for plant responses to climate change. The comprehensive research conducted at the UTHF contributes significantly to our understanding of hemiboreal forest ecosystems.

Keywords Hemiboreal mixed forest · Elevational gradient · Forest dynamics · Long-term ecological research

S. N. Suzuki (✉)
The University of Tokyo Hokkaido Forest, Graduate School of Agricultural and Life Sciences, The University of Tokyo, Furano, Japan

Nakagawa Experimental Forest, Field Science Center for Northern Biosphere, Hokkaido University, Otoineppu, Japan
e-mail: snsuzuki@fsc.hokudai.ac.jp

S. Goto
The University of Tokyo Tanashi Forest, Graduate School of Agricultural and Life Sciences, The University of Tokyo, Nishitokyo, Japan
e-mail: gotos@uf.a.u-tokyo.ac.jp

T. Owari et al. (eds.), *Integrative Forest Management and Silviculture*, Ecological Research Monographs, https://doi.org/10.1007/978-981-95-1017-7_5

5.1 Vegetation of the UTHF

5.1.1 Vegetation Zone: Hemiboreal Mixed Forest Zone

The UTHF is in a hemiboreal mixed forest zone that is primarily composed of boreal conifers and cool-temperate deciduous broadleaves. This type of forest is widespread globally in transitional vegetation zones between cool temperate and boreal forests. Tatewaki (1955, 1958) named this type of forest zone the "pan-mixed forest zone." The major species constituting the hemiboreal forests in Europe, North America, and East Asia, such as spruce (*Picea*), fir (*Abies*), oak (*Quercus*), birch (*Betula*), linden (*Tilia*), maple (*Acer*), and ash (*Fraxinus*), are largely similar at the genus level. In Hokkaido, the mixed forest was composed of evergreen conifers such as *Abies sachalinensis*, *Picea jezoensis*, and *Picea glehnii* and deciduous broadleaves such as *Tilia japonica*, *Acer pictum*, *Betula* spp., and *Quercus crispula*.

In the UTHF, 894 species/subspecies/varieties of vascular plants, including 169 tree species and 78 fern species, have been recorded (Horie et al. 2013). A total of 9.8% of them are endemic to Japan, and 10.6% are alien species. The vegetation is composed of a mix of northern (boreal) and southern (temperate) components of plant communities. The northern components are species that have migrated from the north through Sakhalin and Kuril Islands and are characterized by species such as *Picea jezoensis*, *P. glehnii*, *Urtica angustifolia*, *Crataegus chlorosarca*, *Lonicera alpigena* subsp. *glehnii*, and *Rhododendron dauricum*. The southern components are species that have migrated from the southern direction, that is, mainland Japan, characterized by species such as *Styrax obassia*, *Daphniphyllium macropodum* subsp. *humile*, *Symplocos sawafutagi*, *Vaccinium oldhamii*, and *Arachniodes standishii* (Horie et al. 2013).

5.1.2 Vertical Distribution of Vegetation

As the hemiboreal mixed forest zone is an ecotone, the forest composition changes with elevation even within the UTHF (Kato 1952; Irikura et al. 1979, Figs. 5.1, 5.2, and 5.3). In the lower area (approximately lower than 650 m a.s.l. in elevation), cool-temperate deciduous broadleaves are dominant, but their dominance decreases with elevation. Instead, the dominance of boreal evergreen conifers increases with elevation and reaches a maximum at approximately 800–900 m a.s.l. in elevation. The abundance of *Betula ermanii*, a boreal deciduous broadleaf species, also increases with elevation and should become more dominant in the higher part of the forest zone (Fig. 5.1). Soon below the tree line, sparse *Betula ermanii* forest is dominant. Above the tree line, at approximately 1250 m a.s.l. in the UTHF, dwarf stone pine *Pinus pumila* shrubs are dominant. This elevation of the tree line is ca. 300 m lower than that of the surrounding mountains, such as the Daisetu Mountains, probably due to harsh environments and shallow soils due to mountain ridge

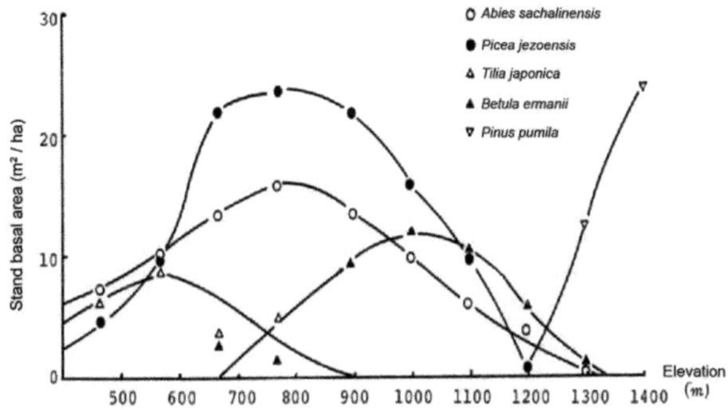

Fig. 5.1 Elevational patterns of the stand basal area of major subalpine species at the University of Tokyo Hokkaido Forest. (Modified after Irikura et al. 1979 by Satoshi N. Suzuki, with the publisher's permission)

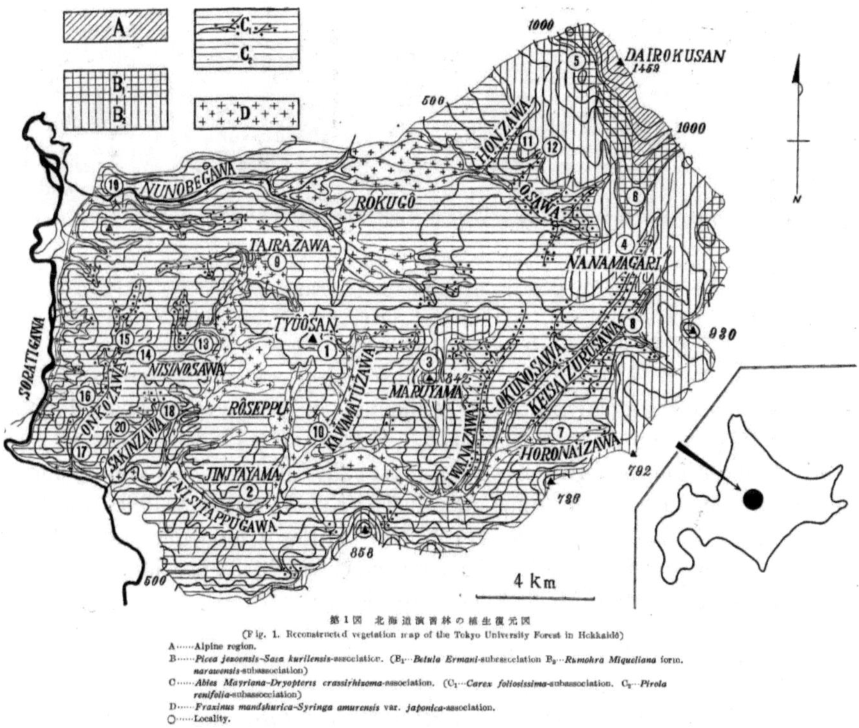

Fig. 5.2 Reconstructed vegetation map of the University of Tokyo Hokkaido Forest. (Reproduced from Kato 1952, with the publisher's permission)

topography (Irikura 1979, Nakata et al. 1994). Above the tree line, there are a variety of alpine plant species, such as *Vaccinium ovalifolium*, *Vaccinium smallii*, *Vaccinium vitis-idaea* (cowberry), *Gaultheria adenothrix*, and *Linneae borealis*.

The forest floor is mostly covered by *Sasa* dwarf bamboos, primarily *Sasa senanensis* (Fig. 5.4), but *Sasa kurilensis* occurs at higher elevations (above ca. 700–800 m asl.). As Sasa dwarf bamboos densely cover forest floors, tree

Fig. 5.3 (**a**) Alpine vegetation (1450 m a.s.l., SEP 7, 2022), (**b**) upper subalpine forest (1230 m a.s.l., JUN 8, 2020), (**c**) lower subalpine forest (700 m a.s.l., AUG 30, 2021), (**d**) conifer-broadleaf mixed forest (340 m a.s.l, Oct. 17, 2009). (**a–c**, photo by Satoshi N. Suzuki; **d**, photo by the UTHF for **d**, licensed under CC BY-NC-ND 4.0)

regeneration is often greatly restricted by them in cool temperate and boreal forests in Japan (Nakashizuka and Numata 1982, Nakashizuka 1988, Kubota and Hara 1996). In these forests, fallen logs or stumps, called nursery logs, play important roles as regeneration sites for some tree species (Kubota and Hara 1996, Narukawa et al. 2003). Additionally, *Sasa* density is relatively low in moderately shaded environments under evergreen conifers (Tatsumi and Owari 2013), providing opportunities for the regeneration of shade-tolerant species.

5.1.3 Vegetation Related to Disturbance, Topography, and Geography

Pioneer species such as birches (*Betula* spp.) often dominate in disturbed areas. The stands that regenerated naturally after fires in the 1910s are now dominated by *Betula maximowicziana* and *Betula platyphylla* var. *japonica*, as well as *Quercus crispula* (Fig. 5.5, see also Chap. 13). On the other hand, the stands that regenerated naturally after severe typhoons, such as those in 1954 and 1981, were dominated by not only *Betula* but also *Abies sachalinensis* and a variety of broadleaves (Fig. 5.6).

a

b

Fig. 5.4 Forest floor covered with *Sasa senanensis* (**a**, August 30, 2021) and seedlings of conifers on a nursery log (**b**, May 29, 2024). (Photo by Satoshi N. Suzuki)

Fig. 5.5 Secondary postfire broadleaf forest dominated by *Betula maximowicziana* after the fire in 1911. (Photo by the UTHF, licensed under CC BY-NC-ND 4.0)

This is likely because advanced regeneration or sprouting from stumps or downed trees contributed to the recovery of the canopy (refer to Chap. 13).

In lowland areas, wetland forests are dominated by species such as *Fraxinus mandshurica* var. *japonica, Alnus japonica*, and *Ulmus davidiana* var. *japonica*. Before the colonization of Japanese people from mainland Japan, it is believed that lowland wet forests, characterized by alder trees, were widely distributed in the Furano Basin and the lowlands of the original UTHF area (Fig. 5.2). However, following the colonization, most of the lowland area was replaced by farmlands, leaving only small remnants in Furano. The largest remnant in Furano was found within the UTHF, where it is now conserved as the Tairazawa Forest Reserve (Fig. 5.7).

Fig. 5.6 A secondary forest regenerated after the typhoon in 1981. (Photo by Satoshi N. Suzuki)

Fig. 5.7 Lowland wet forest in the Tairazawa Forest Reserve. (Photo by Satoshi N. Suzuki, July 1, 2024)

Fig. 5.8 Sakhalin spruce wetland forest in the Maeyama Forest Reserve. (Photo by Satoshi N Suzuki, July 6, 2020)

Some endangered wetland shrubs and trees, such as *Crataegus chlorosarca* and *Padus avium,* grow in the reserved forest.

Picea glehnii often predominates in wetlands at high elevations, approximately 600 m a.s.l. (Fig. 5.8). In such a cold wetland forest, no tree species other than *P. glehnii* can invade the environment because of the overhumidity and unstable nature of the soil (Matsuda 1989).

As limestones are found in the southern part of the UTHF (refer to Chap. 3), limestone vegetation has developed in some limited areas. In the Fujimori Special Plant Community Protection Area, where limestone is exposed to the surface (Fig. 5.9a), limestone plants such as *Berberis amurensis*, *Cortusa matthioli* subsp. *pekinensis* (Fig. 5.9b), *Thalictrum foetidum* var. *glabrescens*, and *Asplenium viride* have been found.

The Chikushimori Special Plant Community Protection Area is on a rocky peak (584 m a.s.l.) of Quarts diabase and is located in the westernmost part of the UTHF. Here, several rocky habitat species, such as *Juniperus chinensis* var. *sargentii*, *Spiraea media* var. *sericea* and *Alnus maximowiczii*, were found. A relatively large population of *Rhododendron albrechtii* was found on a rocky steep slope in the Iwaya Special Plant Community Protection Area.

Fig. 5.9 Exposure of limestone (**a**) and *Cortusa matthioli* subsp. *pekinensis* (**b**) in the Fujimori Special Plant Community Protection Area. (Photo by Kota Kimura, July 3, 2019. Used with permission from the UTHF, licensed under CC BY-NC-ND 4.0)

There are also serpentine areas in the western parts of the UTHF. However, as most of the areas had been mined to extract asbestos, the original vegetation had already been lost. Only a few serpentine species, such as *Viola sachalinensis* var. *alpina* and *Picris hieracioides* subsp. *japonica*, can be found within the UTHF.

Several blowing caves with cold and warm air vents have been found in the UTHF. In these caves, winds blow directionally due to differences in temperature and air pressure, producing air currents that flow in or out of the cave. In the UTHF caves, winds warmer than the ambient air temperature blow out from the vents in the upper positions of slopes in winter, while winds cooler than the ambient air temperature blow out from the vents in the lower positions in summer. Due to the cool winds in summer, several alpine plant species, such as *Vaccinium vitis-idaea* (cowberry), grow around the lower vents even at lower elevations (Matsui and Iguchi 2001, Fig. 5.10).

5.2 Long-Term Research Plots

The UTHF has established forest plots for monitoring the dynamics of natural forests. Most representatives are the Maeyama long-term ecological research (LTER) plot and the Iwanazawa LTER plot.

Fig. 5.10 *Vaccinium vitis-idaea* population around the lower vents of the blowing cave in the Nunobe Special Plant Community Protection Area (340 m a.s.l.). (Photo by Satoshi N. Suzuki, September 6, 2024)

5.2.1 Maeyama LTER Plot

The Maeyama LTER plot spans an area of 36.25 hectares (ha) and is situated on the western slope of Mt. Dairoku, ranging between 620 and 680 meters a. s. l. It is located within the Maeyama Forest Reserve, where light selection cutting was carried out once or twice by 1937. However, since then, no logging or human disturbances have taken place, allowing the forest to reach the state of an old-growth natural forest. The plot was established in 1992 and divided into 145 quadrats measuring 50 m by 50 m (Fig. 5.11, Shibano et al. 1996). Measurements of tree diameter at breast height (DBH) have been conducted at 5- or 10-year intervals for trees with a DBH of 5 cm or more (Ogawa et al. 2021a). These measurements were carried out for all quadrats in 1992–1993, 1997–1998, 2007–2008, and 2017–2018 and for 40 of 145 quadrats (10 ha) in 2002, 2012, and 2022. This plot is the largest among those where trees with a DBH of 5 cm or larger have been consistently measured in natural forests in Japan.

The canopy is dominated by boreal conifers, *A. sachalinensis* and *P. jezoensis*, mixed with deciduous broadleaves such as *Tilia japonica*, *Betula ermanii*, and *Ulmus laciniata* (Fig. 5.12). The understory is predominantly occupied by *Sasa senanensis*, with a few patches of mosses and sedges (*Carex* spp.).

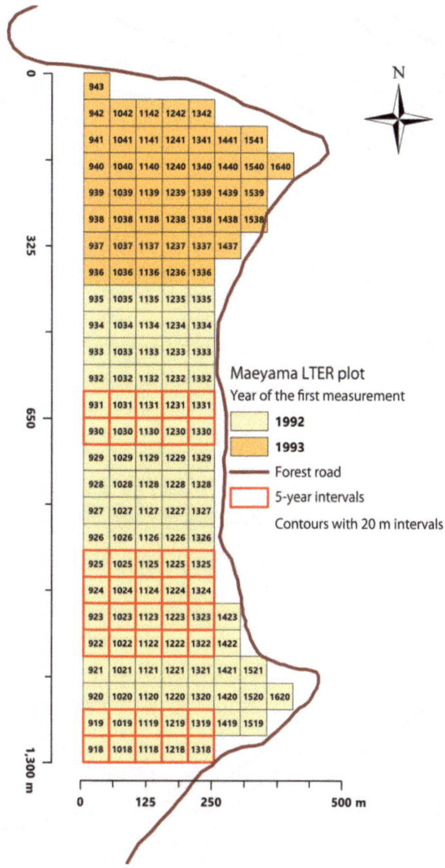

Fig. 5.11 The layout of the Maeyama LTER plot. The number of each square indicates the quadrat ID. (Redrawn from Ogawa et al. 2021a, with the publisher's permission)

The relative dominance of deciduous broadleaves has increased in the plot since the beginning of observation (Fig. 5.13a, Ogawa et al. 2021b). This is mainly because deciduous broadleaves has increased, while conifers have not decreased (Fig. 5.13b). In particular, the relative dominance of deciduous broadleaves was more likely to increase in stands with greater biomass (Fig. 5.13b). A similar increase in deciduous broadleaves has also been reported in other boreal forests in Japan (Suzuki et al. 2015, Hiura et al. 2019) and Canada (Searle & Chen 2017, Hisano et al. 2021). These previous studies suggested that disturbances such as windstorms, climate changes and atmospheric CO_2 enrichment contributed to the increase in deciduous broadleaves (Hiura et al. 2019, Searle & Chen 2017).

The contribution of downed logs to the biodiversity of understory vegetation has been evaluated in the Maeyama LTER (Okada et al. 2019). Not only seedlings of tree species such as *Abies sachalinensis*, *Picea jezoensis*, and *Ulmus laciniate* but also herbaceous species such as *Oxalis acetosella, Circaea alpine, Dryopteris crassirhizoma, and Tiarella polyphylla* were predominantly found on the downed logs. The species compositions of the tree seedlings and herbaceous plants

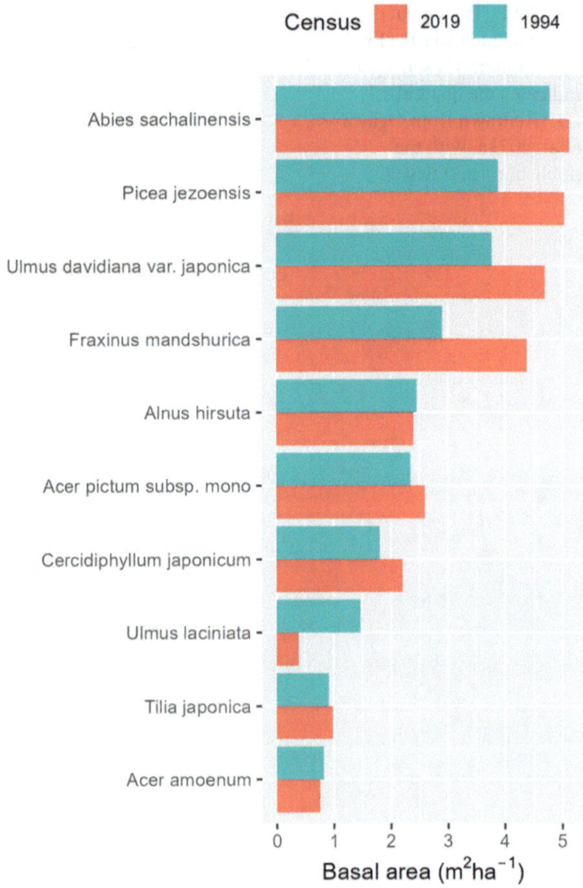

Fig. 5.12 Basal area of major species at the initial (1992–1993) and most recent (2017–2018) measurements in the Maeyama LTER plot. (Redrawn from Ogawa et al. 2021b, with the publisher's permission)

significantly differed between the forest floor and fallen logs and significantly differed between the study quadrats within the habitats (forest floor or downed logs), which contributed to the diversity of the understory in the forest.

5.2.2 *Iwanazawa LTER Plot*

The Iwanazawa LTER plot is in a riparian forest along the Iwanazawa River. It spans 18.75 ha, is 3200 m from south to north, and ranges from 348 to 411 m a. s. l. In this area, a forest tramway was used until the 1950s, during which time weak selection cutting was carried out. The plot was established in 1994 and divided into 75 quadrats measuring 50 m by 50 m (Fig. 5.14). Measurements of the DBH of trees were carried out for all quadrats in 1994, 1999, 2009, and 2019 and for 21 of the 75 quadrats in 2004 and 2014 (Ogawa et al. 2022a). Therefore, the survey was performed every five years for the 5.25-ha subsamples and every 10 years for the remaining subsamples.

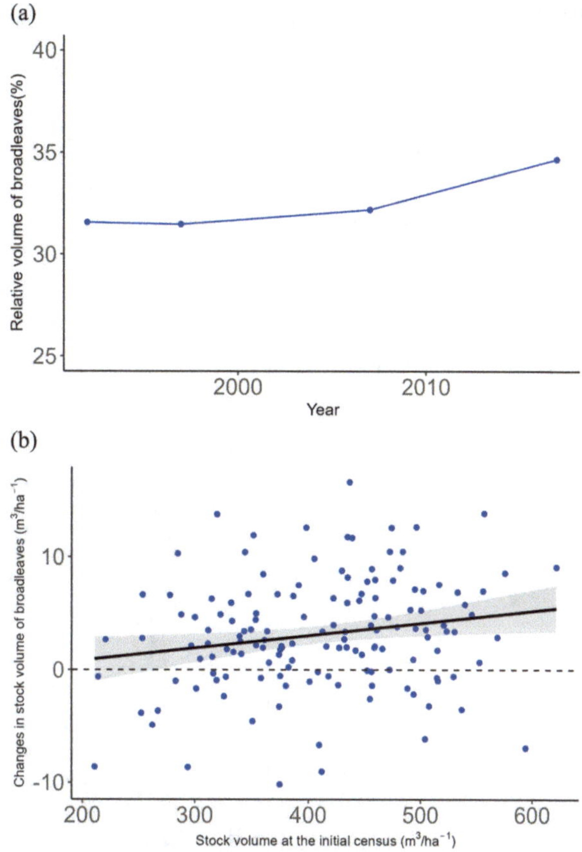

Fig. 5.13 Changes in the relative stock volume of broadleaves (**a**) and the relationship between changes in the stock volume of broadleaves and the total stock volume at the initial census in each quadrat (**b**) in the Maeyama LTER. (Redrawn from Ogawa et al. 2021b, with the publisher's permission)

The canopy is dominated by boreal conifers, *A. sachalinensis* and *P. jezoensis* (10.3%), and riparian broadleaved species such as *U. davidiana* var. *japonica*, *F. mandshurica* var. *japonica*, and *Alnus hirtus* (Fig. 5.15, Ogawa et al. 2022b). In particular, light-demanding species such as *U. davidiana* var. *japonica* and *A. hirtus* were more abundant in areas where forest tramway was installed than in intact areas (Ogawa et al. 2022b).

Some tree species in this riparian area, such as *Ulmus laciniata*, *Hydrangea paniculata*, and *Taxus cuspidata*, have been severely damaged by deer (Ogawa et al. 2022b, refer to also Chap. 14). The frequency of stems damaged by deer was greatest in the period between 1999 and 2009, when more than half of the stems of *U. laciniata*, *H. paniculata*, and *T. cuspidata* were bark-stripped. Due to deer damage, the numbers of live stems of *U. laciniata* and *H. paniculata* steeply decreased until 2009 (Fig. 5.16). Although deer abundance in the UTHF did not decrease (refer to Chap. 14), the frequency of deer damage greatly decreased in the following period between 2009 and 2019 (Fig. 5.17). This can be partly attributed to the decrease in the density of these species palatable to deer and may also be due to changes in the behavior and habitat of deer.

In the Iwanazawa Forest Reserve, which includes the Iwanazawa LTER plot, researches on the dispersal patterns of pollen and seeds of species such as *F.*

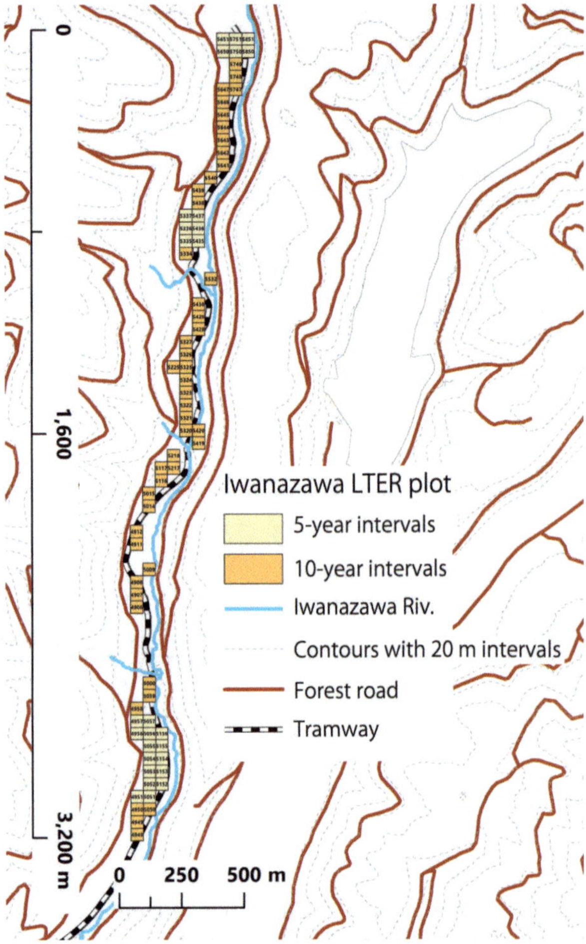

Fig. 5.14 The layout of the Iwanazawa Long-Term Ecological Research Plot. The number in each square represents the quadrat ID. The term "tramway" indicates the rail trace detected in the digital elevation model computed from aircraft LiDAR. The tramway was removed in the 1950s. (Redrawn from Ogawa et al. 2022a, with the publisher's permission)

mandshurica var. *japonica* (Goto et al. 2006), *Cercidiphyllum japonicum* (Nakanishi et al. 2021), Todo fir *A. sachariensis* and *P. jezoensis* (Lian et al. 2008), *Juglans mandshurica* var. *sachalinensis* (Kimura et al. 2012), as well as studies on the genetic structure of species such as *F. mandshurica* var. *Japonica* (Goto et al. 2006), *C. japonicum* (Sato et al. 2006), and *P. jezoensis* (Kitamura et al. 2018), have been actively conducted using a molecular ecological approach. These studies revealed that long-distance dispersal, exceeding 1 km, frequently occurs among these riparian tree species (Fig. 5.18) and that landscape-scale genetic exchange between riparian and adjacent nonriparian forests occurs, indicating that gene flow from external sources plays an important role in maintaining genetic diversity. Considering these points, to conserve riparian forests effectively, it is essential to protect not only the riparian forests themselves but also the surrounding forests.

Based on 6 years of observations of the survival of *P. jezoensis* and *A. sachalinensis plants*, Okada et al. (2015) reported that genetic diversity significantly affected the survival of plants on the forest floor in the Iwanazawa Forest Reserve. They

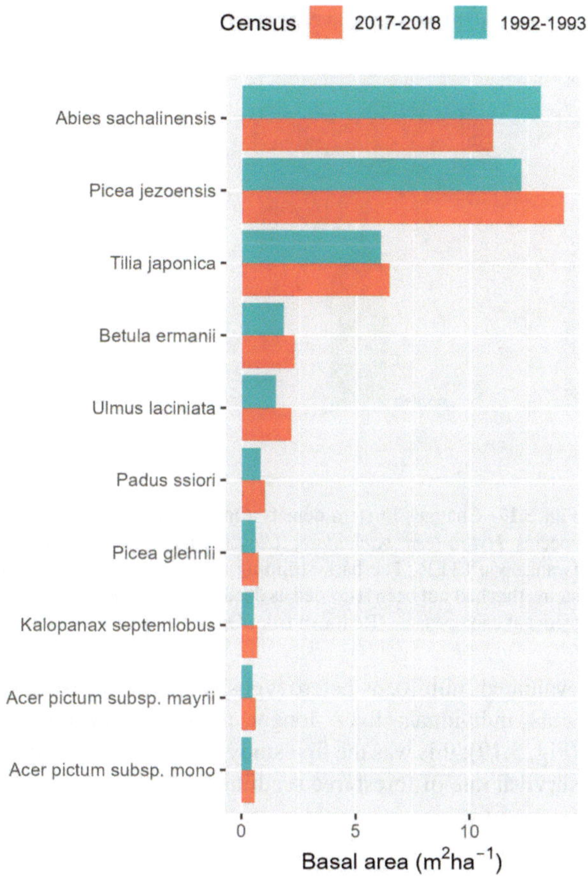

Fig. 5.15 Basal area of major species at the initial (1994) and latest (2019) measurements in the Iwanazawa LTER plot. (Redrawn from Ogawa et al. 2022b, with the publisher's permission)

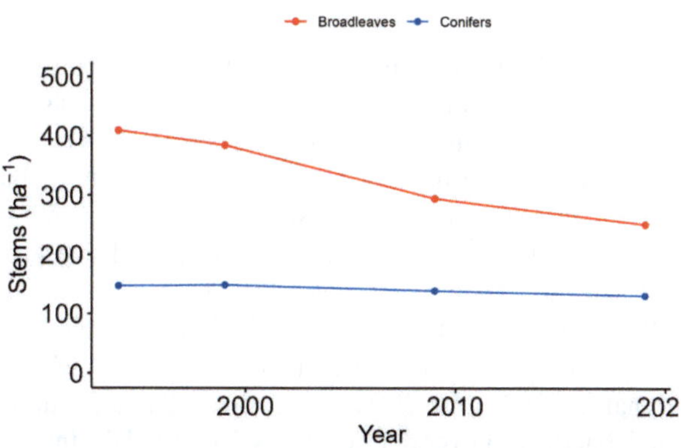

Fig. 5.16 Changes in stem density of conifers (blue) and broadleaves (red) in the Iwanazawa LTER plot. (Data source: UTHF)

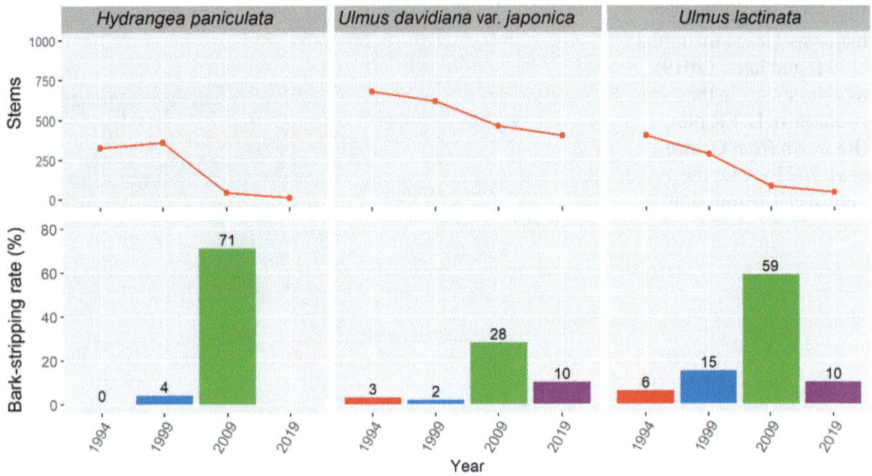

Fig. 5.17 Changes in stem density (upper) and bark-stripping rate (lower) of the deer-favoring species *Hydrangea paniculate*, *Ulmus davidiana* var. *japonica*, and *Ulmus laciniata* in the Iwanazawa LTER. The bark-stripping rate was calculated as the proportion of newly debarked stems that had not been recorded as debarked at previous census, relative to the total number of live stems at each census. (Redrawn from Ogawa et al. 2022b, with the publisher's permission)

evaluated multilocus heterozygosity (MLH) of seedlings and found that heterozygous individuals lived longer than homozygous individuals for *P. jezoensis* (Fig. 5.19); this was the first study to directly quantify the effects of the MLH on the survival rate of forest tree seedlings.

5.2.3 Natural Forest Management Experimental Plots

Since 1955, the UTHF has also established multiple smaller long-term research plots, currently referred to as Natural Forest Management Experimental Plots. The plot size is typically 0.25 ha but ranges from 0.04 to 2.25 ha. These plots were established for the purpose of evaluating the effects of harvesting and estimating the growth rates of forest stands (see also Chap. 8) and have been utilized for a variety of studies. The stand growth, stem growth, and size structure of natural forests under the selection system have been evaluated (Ishibashi and Hirokawa 1986, Ishibashi 1986, Hirata et al. 1987, Hirata and Ishibashi 1987, Tatsuhara et al. 1995, Fig. 5.20). Kitabatake et al. (2003) evaluated the effects of selection cutting on the population dynamics of *A. sachalinensis* and reported that the number of small stems decreased and that the turnover rates of stems decreased during 30 years of dynamics under the selection system. Tatsumi and Owari (2013) and Tatsumi et al. (2016) developed individual-based growth modeling considering the effects of neighborhood competition for predicting long-term effects of selection cutting on forest structure.

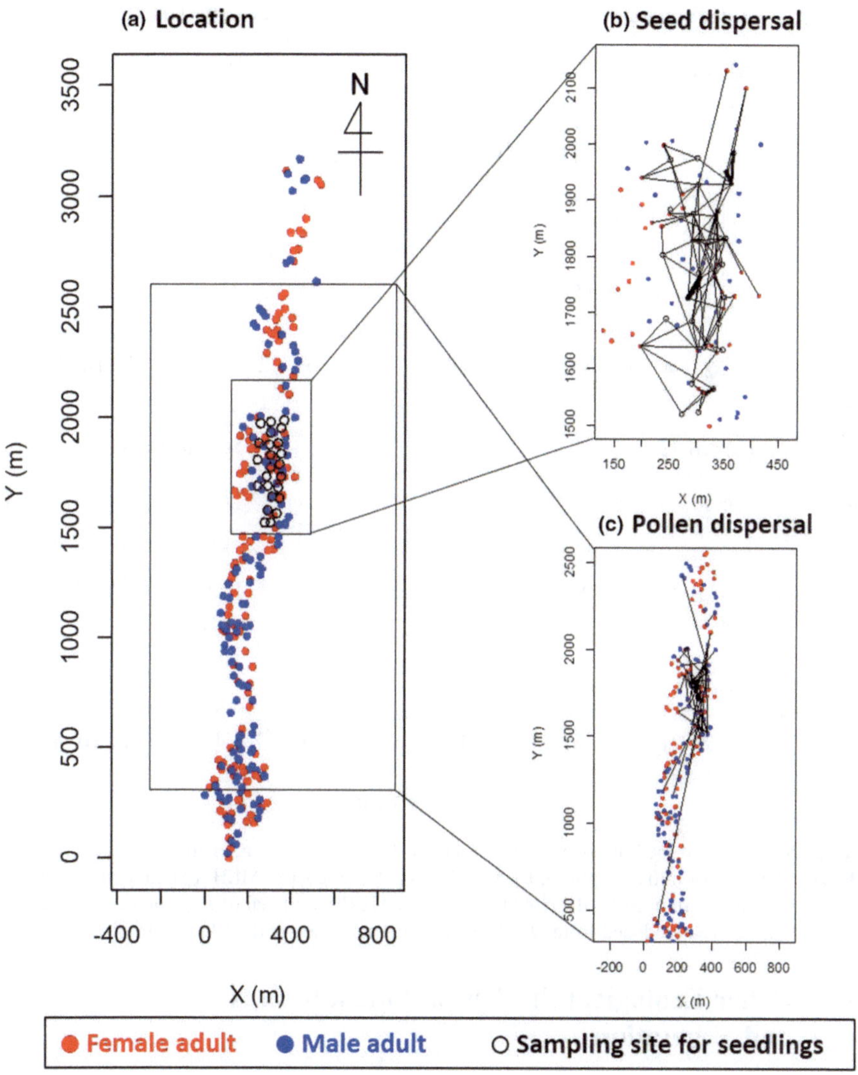

Fig. 5.18 Pollen and seed dispersal of *Cercidiphyllum japonicum* in the Iwanazawa Forest Reserve. (**a**) Locations of the female (blue dots) and male (red dots) adults surveyed and sampling sites for the seedlings (open circles). (**b**) Seed and (**c**) pollen dispersal estimated by genetic analysis. The segments indicate the detected dispersals, and the thickness of each segment increases with increasing frequency of dispersal. (Reproduced from Nakanishi et al. 2021, published under CC BY 4.0)

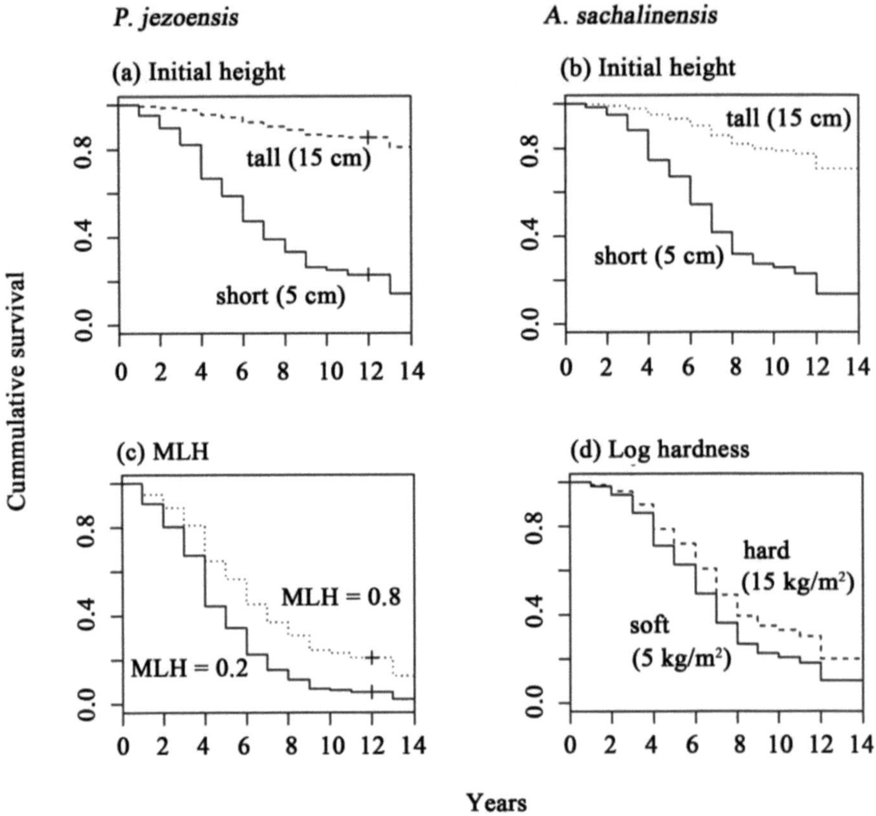

Fig. 5.19 Estimated seedling survival functions for *Picea jezoensis* (**a**, **c**) and *Abies sachalinensis* (**b**, **d**) and effects of initial height (**a**, **b**), multilocus heterozygosity (MLH, **c**) and hardness of logs where seedlings were present (**d**). Note that a higher MLH individual of *P. jezoensis* has a longer longevity in (**c**). (Reproduced from Okada et al. 2015, published under CC BY 4.0)

5.3 Other Ecological Studies on Forest Stands and Vegetation

5.3.1 Biodiversity Experiment

In recent decades, biodiversity-ecosystem function (BEF) has been vigorously studied worldwide via both experimental and field approaches (Ali 2023). In the typical experimental approach, ecosystem functions such as productivity are compared between study quadrats with different numbers of planted species. Initial experimental studies mostly used only herbaceous species due to the feasibility of experimental treatments, but several large-scale experimental studies using tree species started in the 2000s (https://bef-china.com/refer to TreeDivNet, https://treedivnet.ugent.be/). Prior to these studies, a small-scale biodiversity experiment was initiated

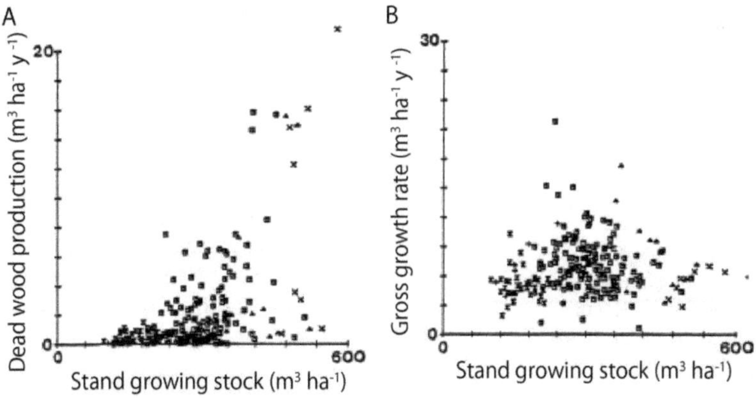

Fig. 5.20 Relationships between the stand growing stock and dead wood production (**a**) and between the stand growing stock and gross growth rate (**b**) in the Natural Forest Management Experimental plots. (Modified after Ishibashi and Hirokawa 1986 by Satoshi N. Suzuki, with the publisher's permission)

at the UTHF in 1987 (Tatsumi 2020). Mixed stands of three species (pioneer *Betula*, mid-successional *Q. crispula*, and shade tolerant, *A. sachalinensis*) and monocultures of each of the three species were set up. At 31 years after planting, the total aboveground biomass of the three-species mixed stands was significantly greater than the average biomass of the monocultures of the three species (Fig. 5.21) and increasing complementarity effects as stand development contributed to the diversity effects of the mixture (Tatsumi 2020). This finding indicates that mixed forests are more efficient at producing woods of diverse species than a combination of monocultures of multiple species. Recently, Tatsumi and Loreau (2023) developed an analytical method for partitioning biodiversity effects into "density-mediated effects" and "size-mediated effects" and found that size-mediated effects (i.e., growth of trees) contributed to net diversity effects at early stages (≤3 years), but density-mediated diversity effects (i.e., mortality) mostly contributed at later stages. This indicates that mixing promotes tree growth via canopy packing in the young stage and that reduced mortality via the complementary use of canopy space enhances stand growth in mixtures.

5.3.2 Elevational Adaptation Studies

Transplanting and common garden experiments are two of the most popular and effective tools for assessing the local adaptation of plant species. Especially, a reciprocal transplant experiment is a powerful tool for detecting local adaptation (Kawecki and Ebert 2004); nevertheless, very few studies have been conducted in conifers. In the UTHF, several transplanting and provenance experiments, including reciprocal transplanting, have been conducted to evaluate the genetic adaptations of

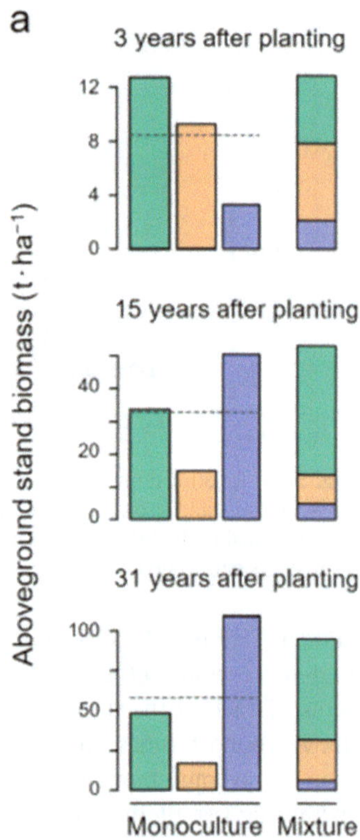

Fig. 5.21 Aboveground stand biomass in the monoculture and mixed stands of *Betula maximowicziana*, *Quercus crispula*, and *Abies sachalinensis* at 3, 15, and 31 years after planting. The horizontal dashed lines indicate the mean values across the monocultures. The difference between the mixture and the mean value of monocultures can be considered the "diversity effect." (Reproduced from Tatsumi 2020, published under CC BY 4.0)

local populations of forest tree species to elevational gradients within the UTHF. The most studied experimental system is the transplanting experiment of *A. sachalinensis*.

The environment, such as temperature, snowfall, and wind speed, changes markedly with elevation. Elevational variation in the growth and morphology of *A. sachalinensis* seedlings was of early interest to researchers. In common garden experiments in which seedlings from high- and low-elevation populations were grown simultaneously in a mild (low-elevation) environment, individuals from high-elevation populations often ceased growing earlier, resulting in smaller seedlings (Oleksyn et al. 1998). While such patterns are commonly reported for conifers, these results alone might suggest that seedlings from lower elevations grow better in any environment. To determine whether this was due to the local adaptation, the UTHF initiated a reciprocal transplant experiment in the 1970s. Open-pollinated seeds of *A. sachalinensis* were collected from eight elevational populations (230 m, 340 m, 420 m, 530 m, 730 m, 930 m, 1100 m, and 1200 m) in 1973. After growing in the nursery, seedlings derived from these populations were planted at six sites (230 m, 420 m, 530 m, 730 m, 930 m, and 1100 m) in 1976 (Kurahashi and Hamaya 1981). Then, the effects of population and site on the phenology, mortality, and growth of 1- to 7-year-old plants were evaluated. Seedlings derived from low-elevation populations exhibited better growth at several sites near the native altitude, whereas high-elevation populations exhibited a low growth rate across all sites (Fig. 5.22).

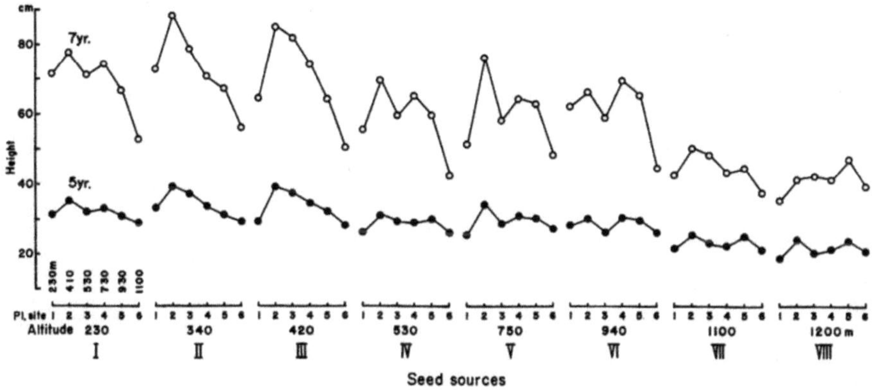

Fig. 5.22 Tree height of 5-year-old (closed circles) and 7-year-old (open circles) *Abies sachalinensis* seedlings in all combinations of seed source elevation (Roman number) and planted site elevation (Arabic number). (Reproduced from Kurahashi and Hamaya 1981, with the publisher's permission)

Furthermore, a mathematical model using data on survival rate and tree height from the early to 36th years at the experimental site confirmed that the average tree height × survival rate decreased with elevational distance from the native site, indicating a native site advantage (Fig. 5.23, Ishizuka and Goto 2012). The results of the 36-year study showed that the low-elevation populations had good growth and survival at low elevations but very low survival at high elevations and that the high-elevation populations had better growth and survival at high elevations but less growth at low elevations than did the low-elevation species. These findings on elevational adaptation in conifers were the result of long-term, large-scale reciprocal transplant experiments conducted in this study.

A high-elevation (1200 m) × low-elevation (530 m) interelevation crossbreeding test site was also created (Kurahashi and Hamaya 1981). The experiment revealed that tree height followed the pattern: high × high < high × low ≈ low × high < low × low, indicating that the reduced growth at high elevations was inherited. For DBH, the results suggested that a reduced vigor, known as distant-cross weakness, occurred in offspring of distantly related crosses, resulting in values lower than the parents' average. (Goto et al. 2011). The number of cones in the study area was counted using an elevated vehicle over a 4-year period, and the results showed that high × high > high × low > low × low > low × high in the fertile years. In other words, reproduction clearly differed between the forward and reverse modes of the cross type, suggesting that there may be a mother-tree effect or epigenetic influence (Hisamoto and Goto 2017). These F_1 individuals are suitable materials for mapping individuals for linkage analysis. Therefore, a mapping population was made by artificial pollination in 2011 using two high × low individuals. The study also examined the genotyping, seedling phenology, and initial growth of the mapping population in the nursery and performed QTL analysis using a linkage map of *A. sachalinensis* with SNP markers via RAD-seq on a next-generation sequencer to identify QTLs

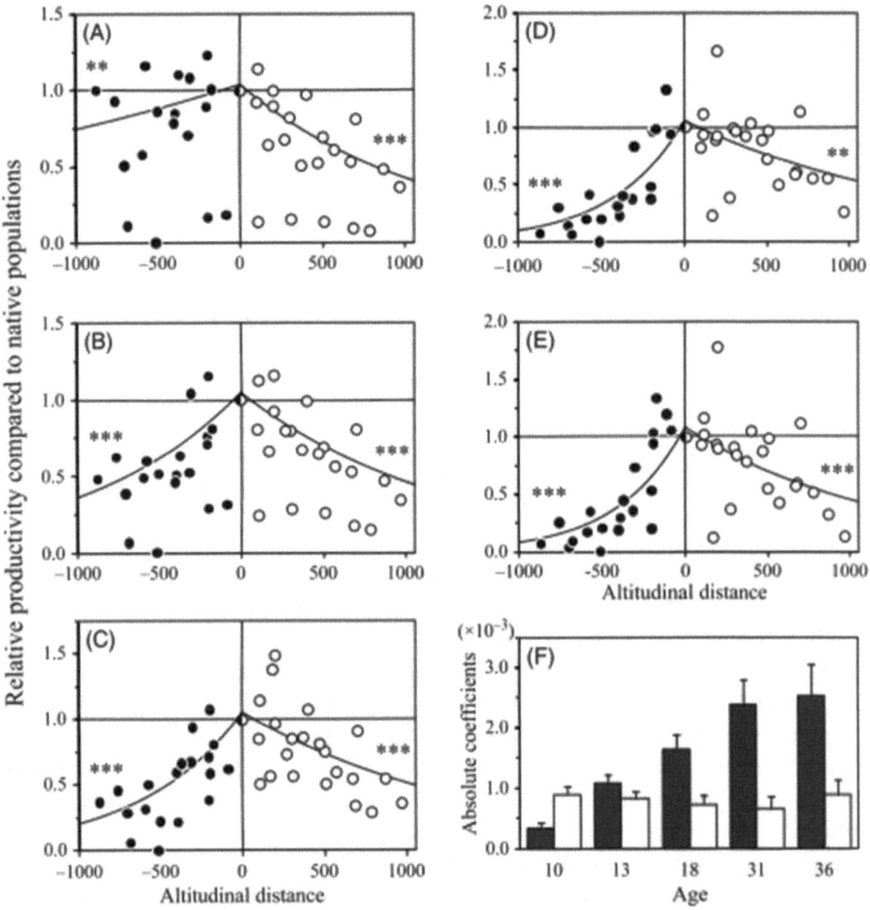

Fig. 5.23 The magnitude of the home-site advantage across the altitudinal range for trees of different ages (**a–e**) and its behavior over time (**f**) in the transplant experiments between different elevation sites. Relative productivities were calculated based on the performance of the transplanted population relative to that of the local population at a given planting site. The plots were arranged in order of the altitudinal distance, DIST, that is, the difference between the elevation of the seed source and the elevation of the planting site, to determine whether the local plants (DIST = 0) outperformed the transplants. The letters **a** to **e** refer to the different testing ages, that is, 10, 13, 18, 31, and 36 years, respectively. A negative DIST value (closed circle) indicates upward transplantation, and positive DIST value (open circle) indicates downward transplantation. Lines indicate the results of regression analyses with the associated significance level, $^{***}P < 0.001$, $^{**}P < 0.01$. The variation in the magnitude of the home-site advantage over time (**f**) is shown in terms of the absolute coefficient of DIST in the regression analysis for trees transplanted upslope (closed bar) and downslope (open bar). The error bars indicate the standard error. A steeper slope reflected more severe transfer effects. (Reproduced from the author's original work, Ishizuka and Goto 2012, published under CC BY-NC)

linked to the trait (Goto et al. 2017). Furthermore, the obtained seedlings were transplanted into long pots, and their photosynthetic and other traits were measured; subsequently, the plants were planted and subjected to linkage analysis for growth traits, and QTLs for Φ_{NO}, one of the photosynthetic parameters, and crown area were detected (Goto et al. 2021).

A common garden experiment of *P. jezoensis* from a range of elevations was also established in 1990. Seeds were collected from 420, 530, 600, 700, 950, 1100, and 1200 m a. s. l. in the fall of 1983 (Nakagawa et al. 2003). Seven-year-old seedlings from all of the elevations were planted at 610 m a.s.l.in elevation, and seedlings from 420, 700, and 1200 m were planted at 750 m a.s.l. in September 1990. Although there was no difference in the timing of bud opening between populations from different elevations in 1997 (Nakagawa et al. 2003), tree height tended to decrease with elevation in 2004 (Fig. 5.24a). The difference was more apparent in 2017 (Fig. 5.24b). This finding implies that boreal conifers, including *A. sachalinensis* and *P. jezoensis,* are also genetically adapted to local elevational conditions, although further studies are needed to assess the nongenetic effects, such as maternal effects or epigenetic variation.

5.3.3 Long-Term Plant Phenology Observations

The UTHF has a long history of phenology observation of budburst, leaf coloration, leaf shedding, and flowering in dozens of tree species (Kamata 2019a). Phenological observations of 43 UTHF species started in 1930 (Nakayama and Kunugi 1940). Although target species have changed over time, the phenology of approximately 30–50 species has been recorded every year except in 1951 (Kamata et al. 2019; Kimura et al. 2022). Unfortunately, the original data during the period 1938–1959 were lost, and only the average values for each species were reported by Kunugi (1952) and Kunugi and Sasaki (1961). As there were some mismatches in the reported values for the same years between the two previous papers, Kamata (2019b) estimated the average values among them.

Long-term data have provided strong evidence of the phenological responses of trees to climate change. While Kimura et al. (1995) reported that the timing of budburst and flowering were significantly correlated with spring temperature, no distinct shift in phenological timing was observed from 1958 to 1993, suggesting that the effects of climate warming were not clear until 1993. However, recent data spanning the last 40 years revealed clear shifts in the phenological timing for nearly all species (Fig. 5.25).

The UTHF maintains a diverse plant community, a result likely influenced by the varied environmental conditions and the establishment of numerous conservation areas (refer also Chap. 15). This vegetational diversity helps to protect a wide range of habitats, contributing to the overall diversity of the ecosystem (refer to also Chap. 6). Furthermore, long-term ecological research at the UTHF has provided valuable insights into the fundamental dynamics of hemiboreal forests, enhancing our

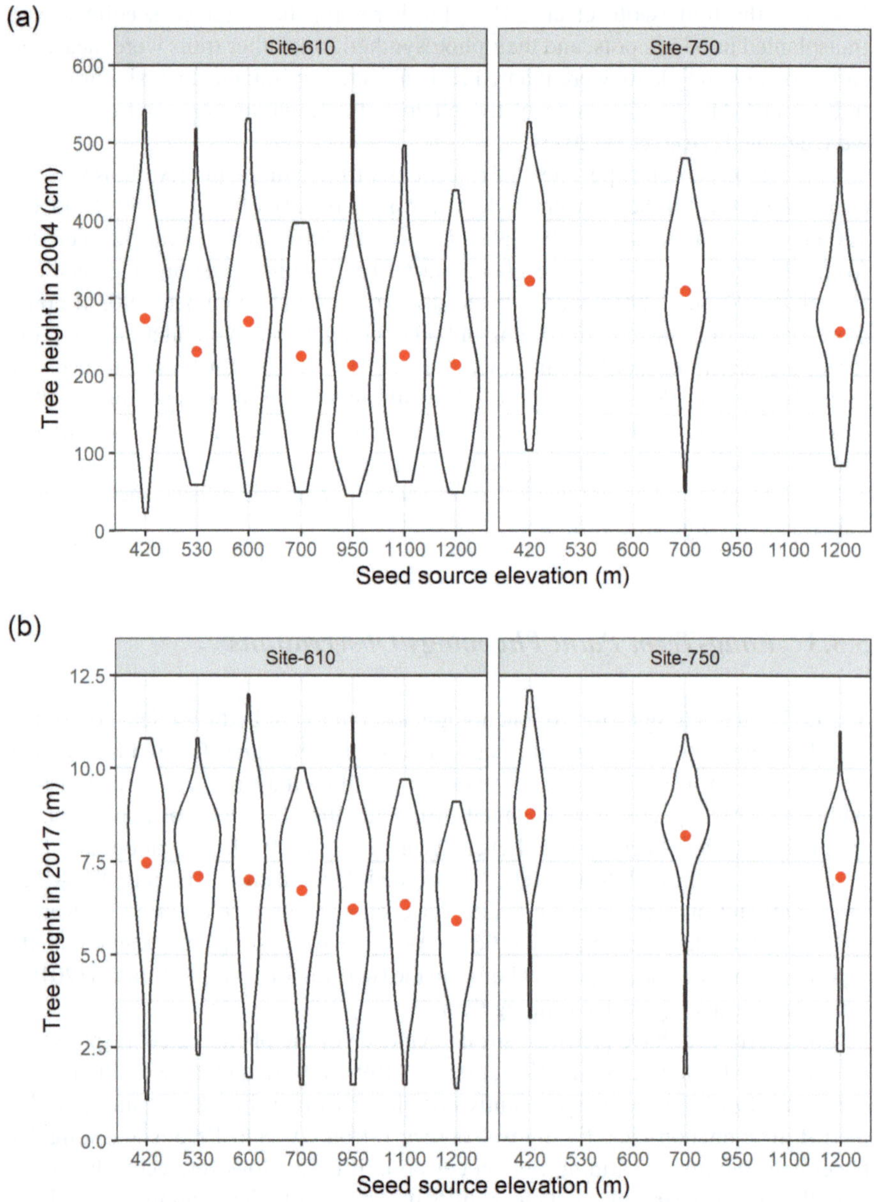

Fig. 5.24 The frequency distribution of tree height of *P. jezoensis* from different elevations at the two sites, measured at 15 years old (**a**) and 28 years old (**b**). The red closed circle indicates the average. (Data source: the UTHF)

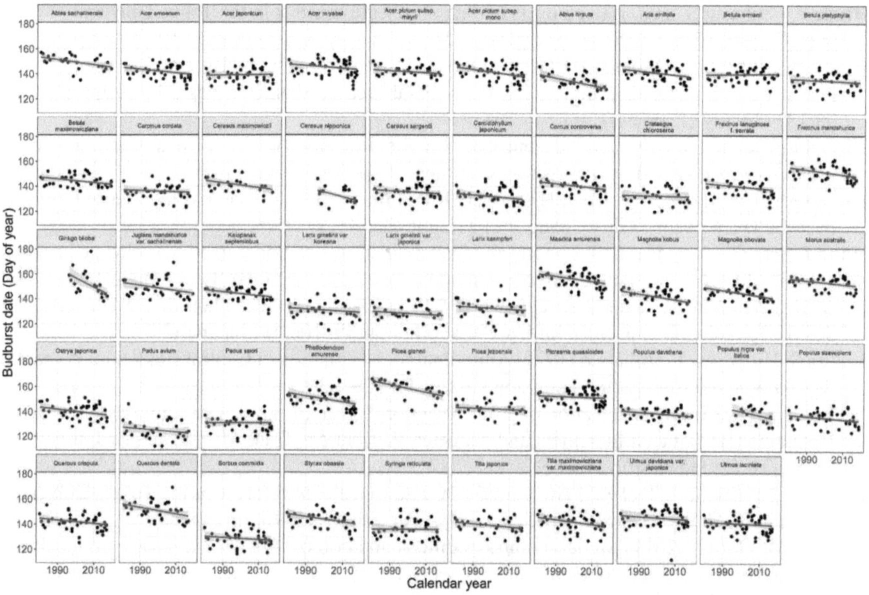

Fig. 5.25 Forty-year observations (1981–2020) of budburst timing of 49 species in the UTHF. (Data compiled from Kamata et al. 2019 and Kimura et al. 2022)

understanding of these complex systems. Based on this extensive body of knowledge, the UTHF has adopted sustainable methods of forest management practices, the Stand-based Silvicultural Management System, that promote the natural regeneration of the forest and maintain ecological balance (refer to Chap. 7). By avoiding extensive clear-cutting, the UTHF ensures that the large portion of the forest remains in a mature or old-growth natural state, which is crucial for the survival of many species and the overall health of the ecosystem.

References

Ali A (2023) Biodiversty-ecosystem functioning research: brief history, major trends and perspectives. Biol Conserv 285:110210

Goto S, Shimatani K, Yoshimaru H, Takahashi Y (2006) Fat-tailed gene flow in the dioecious canopy tree species *Fraxinus mandshurica* var. *japonica* revealed by microsatellites. Mol Ecol 15:2985–2996

Goto S, Iijima H, Ogawa H, Ohya K (2011) Outbreeding depression caused by intraspecific hybridization between local and nonlocal genotypes in *Abies sachalinensis*. Restor Ecol 19:243–250

Goto S, Kajiya-Kanegae H, Ishizuka W, Kitamura K, Ueno S, Hisamoto Y, Kudo H, Yasugi M, Nagano AJ, Iwata H (2017) Genetic mapping of local adaptation along the altitudinal gradient in *Abies sachalinensis*. Tree Genet Genomes 13:1–13

Goto S, Mori H, Uchiyama K, Ishizuka W, Taneda H, Kono M, Kajiya-Kanegae H, Iwata H (2021) Genetic dissection of growth and eco-physiological traits associated with altitudinal adaptation in Sakhalin fir (*Abies sachalinensis*) based on QTL mapping. Genes 12:1110

Hirata Y, Ishibashi S (1987) An analysis of data on selection forest in the Tokyo University Forest in Hokkaido (III) diameter grouping. Trans Meet Kanto Br Jpn For Soc 39:19–22. (in Japanese)
Hirata Y, Ishibashi S, Minowa M (1987) On stand structure of natural forests (I) needle-broad leaved pair diameter distribution and mean basal area by diameter class. Trans Jpn For Soc 98:133–134. (in Japanese)
Hisamoto Y, Goto S (2017) Genetic control of altitudinal variation on female reproduction in *Abies sachalinensis* revealed by a crossing experiment. J For Res 22:195–198
Hisano M, Ryo M, Chen X, Chen HY (2021) Rapid functional shifts across high latitude forests over the last 65 years. Glob Chang Biol 27:3846–3858
Hiura T, Sato G, Iijima H (2019) Long-term forest dynamics in response to climate change in northern mixed forests in Japan: a 38-year individual-based approach. For Ecol Manag 449:117469
Horie K, Miyamoto Y, Kimura N, Oikawa N (2013) List of vascular plant of the University of Tokyo Hokkaido Forest. Misc Inf Univ Tokyo For 54:59–106. (in Japanese)
Irikura S, Hirokawa T, Shibata Z (1979) Altitudinal forest zonation of Central Hokkaido (I) species distribution and forest zone. Trans Jpn For Soc 90:329–330. (in Japanese)
Ishibashi S (1986) An analysis of data on selection forest in the Tokyo University Forest in Hokkaido (II) diameter increment and diameter distribution. Trans Meet Kanto Br Jpn For Soc 38:31–34. (in Japanese)
Ishibashi S, Hirokawa T (1986) An analysis of data on selection forest in the Tokyo University Forest in Hokkaido. Trans Meet Kanto Br Jpn For Soc 37:15–18. (in Japanese)
Ishizuka W, Goto S (2012) Modeling intraspecific adaptation of *Abies sachalinensis* to local altitude and responses to global warming based on a 36-year reciprocal transplant experiment. Evol Appl 5:229–244
Kamata N (2019a) Long-term phenology data of woody plants at the university of Tokyo Hokkaido Forest. In: Kamata N, Kuraji K, Owari T, Guan BT (eds) Developing a network of long-term research field station to monitor environmental changes and ecosystem responses in Asian forests. The University of Tokyo Forests Press, Tokyo
Kamata N (2019b) Estimation of mean values of phenology data on woody plants at the University of Tokyo Hokkaido Forest for 1938–1950, when the original data have been lost. Bull Univ Tokyo For 140:1–13. (in Japanese with English Summary)
Kamata N, Kimura N, Iguchi K, Fukuoka S, Ogawa H, Kasahara H, Shibano S, Takahashi Y, Inukai M, Sasaki C et al (2019) Long-term phenology data on woody plants at the University of Tokyo Hokkaido Forest from 1930 to 2010. Misc Info Univ Tokyo For 61:45–63. (Japanese with English summary)
Kato R (1952) The vegetation of the Tokyo University Forest in Hokkaido. Bull Univ Tokyo For 43:1–18. (in Japanese with English summary)
Kawecki TJ, Ebert D (2004) Conceptual issues in local adaptation. Ecol Lett 7:1225–1241
Kimura MK, Goto S, Suyama Y, Matsui M, Woeste K, Seiwa K (2012) Morph-specific mating patterns in a low-density population of a heterodichogamous tree, *Juglans ailantifolia*. Plant Ecol 213:1477–1487
Kimura K, Kisanuki H, Kurahashi A (1995) Some phenological characters on 25 broad-leaved deciduous tree species in Hokkaido. Annual fluctuation and effect of monthly mean temperature. Trans Jpn For Soc 106: 367–370 (in Japanese)
Kimura N, Fukuoka S, Ogawa H, Hirota Y, Kamata N (2022) Phenology data on woody plants at the University of Tokyo Hokkaido Forest from 2011 to 2020. Misc Info Univ Tokyo For 65:103–110. (Japanese with English summary)
Kitabatake T, Goto S, Takahashi Y, Kasahara H, Inukai M (2003) Effect of selection cutting on population dynamics of *Abies sachalinensis* in a cool temperate mixed forest in Hokkaido, Japan. Jpn J For Soc 85:252–258
Kitamura K, Nakanishi A, Lian C, Goto S (2018) Distinctions in fine-scale spatial genetic structure between growth stages of *Picea jezoensis*. Front Genet 9:Article 490
Kubota Y, Hara T (1996) Recruitment processes and species coexistence in a subboreal forest in northern Japan. Ann Bot 78:741–748

Kurahashi A, Hamaya T (1981) Variation of morphological characteristics and growth response of Saghalien fir (*Abies sachalinensis*) in different altitudes. Bull Tokyo Univ For 71:101–151

Kunugi R (1952) Kisetsu cyousa ni yoru rinboku no heni gensyou to sono ouyousei [Seasonal survey of variation phenomena in forest trees and their applicability]. Trans Meet Hokkaido Br Jpn For Soc 1: 31–37 (in Japanese)

Kunugi R, Sasaki C (1961) Syuyou jyumoku no kisetsu cyousa siryou [Seasonal survey data of major tree species]. Trans Meet Hokkaido Br Jpn For Soc 10: 59–63 (in Japanese)

Lian C, Goto S, Kubo T, Takahashi Y, Nakagawa M, Hogetsu T (2008) Nuclear and chloroplast microsatellite analysis of *Abies sachalinensis* regeneration on fallen logs in a subboreal forest in Hokkaido, Japan. Mol Ecol 17:2948–2962

Matsuda K (1989) Regeneration and growth in the *Picea glehnii* forest. Res Bull Hokkaido Univ For 3:595–717

Matsui M, Iguchi K (2001) Vegetation and change of underground temperature throughout the year in the area of NUNOBE "wind-hole". Trans Meet Hokkaido Br Jpn For Soc 49:76–78

Nakagawa M, Hirokawa T, Shima T, Ogasawara S, Kurahashi A (2003) Effect of seed source elevation on bud opening of Yezo spruce (*Picea jezoensis*). J For Res 8:267–270

Nakanishi A, Goto S, Sumiyoshi C, Isagi Y (2021) Directional seed and pollen dispersal and their separate effects on anisotropy of fine-scale spatial genetic structure among seedlings in a dioecious, wind-pollinated, and wind-dispersed tree species, *Cercidiphyllum japonicum*. Ecol Evol 11:7754–7767. https://doi.org/10.1002/ece3.7609

Nakashizuka T (1988) Regeneration of beech (Fagus crenata) after the simultaneous death of undergrowing dwarf bamboo (*Sasa kurilensis*). Ecol Res 3:21–35

Nakashizuka T, Numata M (1982) Regeneration process of climax beech forests I. Structure of a beech forest with the undergrowth of *Sasa*. Jpn J Ecol 32:57–67

Nakata M, Tanaka H, Yagi H (1994) Altitudinal changes in vegetation and soil on Mt. Dairoku, central Hokkaido, Japan. Jpn J Ecol 44:33–47. (in Japanese with English summary)

Nakayama M, Kunugi R (1940) Jyumoku no kaigaki, kaikaki, kouyou oyobi rakuyou gensyou to kiyou-innshi tono kankei ni tsuite [On the relationship between tree budburst, flowering, autumn foliage, and leaf fall phenomena and meteorological factors]. Misc Info Univ Tokyo For 1: 29–82 (in Japanese)

Narukawa Y, Iida S, Tanouchi H, Abe S, Yamamoto S-I (2003) State of fallen logs and the occurrence of conifer seedlings and saplings in boreal and subalpine old-growth forests in Japan. Ecol Res 18:267–277

Ogawa H, Shibano S, Okamura K, Takahashi Y, Ohkawa A, Kasahara H, Kaji M, Yamamoto H, Suzuki S (2021a) Growth records of the Maeyama long-term ecological research plot in the University of Tokyo Hokkaido Forest (1992–2018). Misc Inf Univ Tokyo For 63:113–123

Ogawa H, Ohkawa A, Kasahara H, Suzuki S (2021b) 25-year dynamics of hemiboreal conifer-broadleaf mixed forest in a 36-ha long term ecological research plot. Boreal For Res 69:31–34

Ogawa H, Iguchi K, Takahashi Y, Okamura K, Ohkawa A, Matsui M, Suzuki SN (2022a) Growth records of trees in the Iwanazawa long-term ecological research plot in the University of Tokyo Hokkaido Forest (1994–2019). Misc Inf Univ Tokyo For 65:71–80

Ogawa H, Ohkawa A, Iguchi K, Suzuki SN (2022b) 25-year dynamics of a riparian forest in an 18 ha long-term ecological research plot. Boreal For Res 70:49–52

Okada M, Kitamura K, Lian C, Goto S (2015) The effects of multilocus heterozygosity on the longevity of seedlings established on fallen logs in *Picea jezoensis* and *Abies sachalinensis*. Open J For 5:422–430

Oleksyn J, Modrzýnski J, Tjoelker MG, Żytkowiak R, Reich PB, Karolewski P (1998) Growth and physiology of *Picea abies* populations from elevational transects: common garden evidence for altitudinal ecotypes and cold adaptation. Funct Ecol 12:573–590

Sato T, Isagi Y, Sakio H, Osumu K, Goto S (2006) Effect of gene flow on spatial genetic structure in the riparian canopy tree *Cercidiphyllum japonicum* revealed by microsatellite analysis. Heredity 96:79–84

Searle EB, Chen HY (2017) Persistent and pervasive compositional shifts of western boreal forest plots in Canada. Glob Chang Biol 23:857–866

Shibano S, Okamura K, Takahashi Y, Watanabe S (1996) Techniques for establishing a large permanent research plot for studies of forest dynamics in a mixed forest zone. Jpn J Ecol 45:155–168. (in Japanese with English summary)
Suzuki SN, Ishihara M, Hidaka A (2015) Regional-scale directional changes in abundance of tree species along a temperature gradient in Japan. Glob Chang Biol 21:3436–3444
Tatewaki M (1955) Pan-mixed forest. North For 7:8–11. (in Japanese)
Tatewaki M (1958) Forest ecology of the islands of the North Pacific Ocean. J Fac Agric Hokkaido Univ 50:371–486
Tatsuhara S, Yamamoto H, Takada N, Inukai H (1995) Effects of stand structure on growth in selection stands dominated by Todo fir (*Abies sachalinensis*). Jpn J For Plann 25:17–25
Tatsumi S (2020) Tree diversity effects on forest productivity increase through time because of spatial partitioning. For Ecosyst 7:24
Tatsumi S, Loreau M (2023) Partitioning the biodiversity effects on productivity into density and size components. Ecol Lett 26:1963–1973
Tatsumi S, Owari T (2013) Modeling the effects of individual-tree size, distance, and species on understory vegetation based on neighborhood analysis. Can J For Res 43:1006–1013
Tatsumi S, Owari T, Mori AS (2016) Estimating competition coefficients in tree communities: a hierarchical Bayesian approach to neighborhood analysis. Ecosphere 7:e01273

Chapter 6
Fauna and Their Biology

Dai Fukui and Naoto Kamata

Abstract The UTHF is home to many animal species. Although it is nearly impossible to cyclopedically determine the insect fauna, numerous intensive studies have been conducted to investigate specific groups of insect fauna. Regarding mammals, 33 out of 53 species present in Hokkaido have been confirmed in the UTHF. Bird and fish fauna have also been studied. To date, a total of five orders, 40 families, 84 genera, and 121 species of birds have been recorded, including three species designated as the natural monuments of Japan. In terms of fish, a total of nine species have been documented, including two introduced species and one hybrid.

Keywords Insects · Mammals · Birds · Fishes

6.1 Introduction

Forests support a wide range of species, and a comprehensive understanding of these species list is fundamental to an effective forest management and scientific research. Since the early twenty-first century, the UTHF has been systematically compiling an inventory of living fauna as part of basic data development, with a notable progress achieved for certain taxonomic groups. However, taxa characterized by high species diversity, such as insects, remain insufficiently documented. This chapter provides an overview of the fauna of insects and vertebrates (mammals, birds, and fishes) recorded in the University of Tokyo Hokkaido Forest so far.

D. Fukui (✉)
The University of Tokyo Fuji Iyashinomori Woodland Study Center, Graduate School of Agricultural and Life Sciences, The University of Tokyo, Yamanakako, Japan
e-mail: fukuidai@uf.a.u-tokyo.ac.jp

N. Kamata
The University of Tokyo Chiba Forest, Graduate School of Agricultural and Life Sciences, The University of Tokyo, Kamogawa, Japan
e-mail: kamatan@uf.a.u-tokyo.ac.jp

T. Owari et al. (eds.), *Integrative Forest Management and Silviculture*,
Ecological Research Monographs, https://doi.org/10.1007/978-981-95-1017-7_6

6.2 Insects

The class Insecta is by far the most prosperous taxonomic group on the globe, including approximately 30,000 species from Japan and approximately one million species in the world, which accounts for more than half of the described species (approximately 1,750,000) (Umeya 2003). Cyclopedic determination of the insect fauna in the UTHF is nearly impossible. However, since the mid-twentieth century, numerous intensive studies have been conducted to investigate specific groups of insect fauna on specific host plants (Table 6.1). Taxonomically, the orders Coleoptera and Lepidoptera have received the most attention (Fig. 6.1). The two most studied guilds are defoliating insects and wood borers including phloem feeders (Fig. 6.2). Recently, fauna of flower longicorn beetles belonging to the subfamilies Lepturinae and Necydalinae (Coleoptera, Cerambycidae) in the UTHF has been documented (Toki et al. 2024) (Fig. 6.3).

Some described species that were not found in Japan or Hokkaido were recorded from the UTHF. *Pachyligia dolosa* (Lepidoptera: Geometridae) collected from the UTHF was the first record from Hokkaido (Arisawa 1965a). Two scolytine beetles, *Scolytoplatypus blandfordi* and *Anisandrus hirtus* (Coleoptera: Curclionidae), captured in the UTHF, represent the first records of each species in Japan (unpublished). It is widely recognized that species diversity tends to decrease with temperature. Therefore, it is speculated that species diversity in Hokkaido is lower than in other parts of Japan. A series of studies on bark beetle (*sensu lato*) fauna suggest that the temperature gradient along latitudinal gradient influenced a gradient of bark beetle diversity both directly and indirectly through the gradient of plant diversity (Peng 2022). The decline in plant diversity along the latitudinal gradient increased the host specificity of the bark beetles and reduced the stochasticity and diversity of the bark beetle fauna. However, a different pattern emerged from the ground beetle survey conducted in the seven university forests of the University of Tokyo Forests using the same methodology (Insect Group 2024). The results showed that the species richness in the UTHF was not lower compared to the other six forests, which are situated further south. For instance, the number of taxa belonging to the family Carabidae (Coleoptera) was the second largest among the seven forests surveyed. A similar trend was observed in a nationwide project called Monitoring Sites 1000, which indicated that species richness and abundance of ground-dwelling beetles tend to be greater in cooler regions (Biodiversity Center of Japan 2009). In these cooler areas, food resources may be more plentiful due to slower decomposition rates, supporting a greater number of species and higher abundances of species within the detritivorous food web, including ground-dwelling beetles (Wallon et al. 2024).

Extensive ecological studies have been conducted on certain pest species. For further details, refer to Chap. 14.

Table 6.1 Number of recorded insect species within each of the three major orders from intensive studies on insect fauna at the University of Tokyo Hokkaido Forest, along with their target host plant

Publication	Insect guild	Insect order			Target host plant			
		COL	LEP	HYM	Order	Family	Genus	Species
Nishiguchi (1957a)	Wood borer	4			Coniferae	Pinaceae	*Pinus*	*strobus*
Nishiguchi (1960d)	Wood borer	16			Coniferae	Pinaceae	*Pinus*	*strobus*
Nishiguchi (1957a)	Wood borer	4			Coniferae	Pinaceae	*Pinus*	*koraiensis*
Nishiguchi (1960d)	Wood borer	18			Coniferae	Pinaceae	*Pinus*	*koraiensis*
Nishiguchi (1957a)	Wood borer	4			Coniferae	Pinaceae	*Pinus*	*sylvestris*
Nishiguchi (1960d)	Wood borer	5			Coniferae	Pinaceae	*Pinus*	*sylvestris*
Nishiguchi (1957a)	Wood borer	1			Coniferae	Pinaceae	*Pinus*	*ponderosa*
Nishiguchi (1960d)	Wood borer	5			Coniferae	Pinaceae	*Pinus*	*ponderosa*
Nishiguchi (1957b)	Wood borer	3			Coniferae	Pinaceae	*Pseudotsuga*	*menziesii*
Nishiguchi (1960d)	Wood borer	2			Coniferae	Pinaceae	*Pseudotsuga*	*menziesii*
Nishiguchi (1959)	Wood borer	37			Coniferae	Pinaceae	*Picea*	*abies*
Nishiguchi (1960c)	Wood borer	15+			Coniferae	Pinaceae	*Picea*	*abies*
Nishiguchi (1958)	Defoliator		11+	1	Malpighiales	Salicaceae	*Populus*	*gigantea*
Nishiguchi (1960a)	Defoliator		31		Malpighiales	Salicaceae	*Populus*	*gigantea*
Nishiguchi (1960b)	Defoliator	15			Malpighiales	Salicaceae	*Populus*	*davidiana*
Nishiguchi (1962a)	Defoliator		23		Malpighiales	Salicaceae	*Populus*	*davidiana*
Nishiguchi (1962b)	Wood borer		1		Malpighiales	Salicaceae	*Populus*	*davidiana*
Kamata et al. (2014)	Wood borer	7			Fagales	Betulaceae	*Betula*	*maximowicziana*
Peng et al. (2022)	Wood borer	9			Fagales	Betulaceae	*Betula*	*platyphylla* var. *japonica*
Peng et al. (2022)	Wood borer	7			Fagales	Betulaceaee	*Alnus*	*hirsuta*
Arisawa (1965b)	Defoliator		74		Miscellaneous			
Toki et al. (2024)	Flower longhorn beetle	33			Miscellaneous			

Fig. 6.1 Insect specimens stored in the backyard of Forest Exhibition Hall, The University of Tokyo Hokkaido Forest. Longicorn beetles (Coleoptera: Cerambycidae) (left) and hawk moths (Lepidoptera: Sphingidae) (right). (Photos by Naoto Kamata)

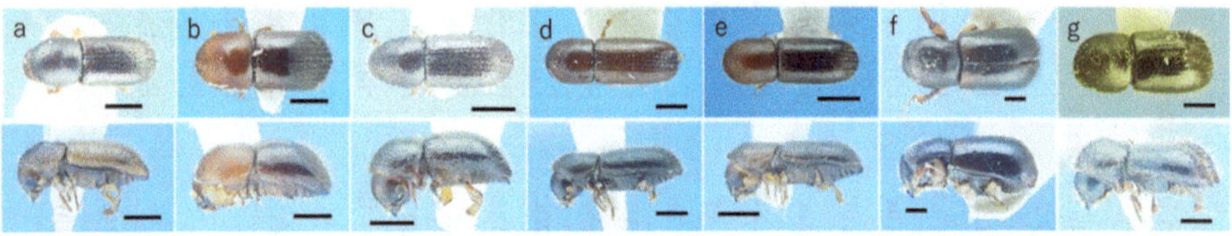

Fig. 6.2 Dorsal (upper) and lateral (lower) photographs of ambrosia beetles (Coloeptera: Curclionidae) captured from *Betula maximowicziana*, which have been severely defoliated for several years. (a) *Heteroborips seriatus* (Blandford), (b) *Xylosandrus crassiusculus* (Motschulsky), (c) *Xylosandrus germanus* (Blandford), (d) *Xyleborinus attenuatus* (Blandford), (e) *Xyleborinus saxesenii* (Ratzeburg), (f) *Scolytoplatypus daimio* Blandford, and (g) *Anisandrus apicalis* (Blandford) (after Kamata et al. (2014). Photos by Naoto Kamata)

Fig. 6.3 *Rhagium heyrovskyi* (Coleoptera: Cerambycidae) on a log of *Picea jezoensis*. (Photo by Wataru Toki, licensed under CC BY-NC-ND 4.0)

6.3 Mammals

Until the beginning of this century, researches on mammals in the UTHF had primarily focused on rodents, hare, and sika deer, species that significantly impact forestry or forest vegetation (refer to Chap. 14). Systematic surveys of mammalian fauna had been nearly absent, with the exception of Shibano and Takeuchi (1983). However, in 2012, the Fundamental Data Development Committee commenced monitoring surveys of medium and large mammals utilizing sensor cameras (Oikawa and Matsui 2017; Oikawa et al. 2022a). In addition, between 2015 and 2020, research was conducted on bat fauna, which are difficult to identify using sensor cameras. Regarding rodents, capturing surveys to infer population density have been conducted continuously since the 1960s to predict forest damage caused by voles. Initially, these surveys were not conducted at fixed locations and detailed records of captures were not maintained. However, since the year 2000, fixed survey sites have been established and capture records have been systematically kept (Akiba et al. 2013). This improvement in methodology has led to a much better understanding of the rodent fauna.

Through a series of research to date, 33 species of mammals in 6 orders and 13 families have been confirmed in the UTHF (Table 6.2, Figs. 6.4, 6.5, 6.6, and 6.7). The most common order is Chiroptera, with 2 families and 11 species, followed by Rodentia. The confirmed mammalian fauna includes feral dogs, feral cats, and four nonnative species (Norway rat, house mouse, raccoon, and American mink). Raccoons, in particular, have been photographed more frequently in forests in recent years, suggesting that their distribution area is expanding (refer to Chap. 15).

Although the information on the population size and dynamics of each species is generally limited, several consecutive surveys have been conducted for some species or taxa. Light census monitoring, initiated in 2007, revealed that the population density of sika deer, *Cervus nippon*, which rapidly increased in the 1990s (Takahashi et al. 1997), has remained consistently high in the UTHF (UTHF unpublished data). For rodents, although fluctuations in the number of captured individuals are observed since 1960s as mentioned above, however, no clear cyclical trends have been detected since 2000, when surveys began at fixed survey sites (UTHF unpublished data). In the rodent survey, two survey sites were set up at high altitudes (550–680 m) and two at low altitudes (410–420 m). There is a tendency for the grey red-backed vole, *Craseomys rufocanus*, to be more prevalent at high altitudes, and the large Japanese field mouse, *Apodemus speciosus*, to be more prevalent at low altitudes. There is no clear difference in the population of the small Japanese field mouse, *Apodemus argenteus*, between the altitudes. Regarding Siberian flying squirrels, *Pteromys volans orii*, it has been reported that there are two individuals per ha in an *Abies sachalinensis*-dominated mixed forest (Suzuki et al. 2011).

In addition to the aforementioned faunal surveys, ecological studies on several species have also been conducted within the UTHF. For the rare Japanese pika, *Ochotona hyperborea yesoensis*, Oikawa and Matsui (2009) revealed that a population of Mt. Dairoku is maintained above an altitude of approximately 1340 m. Oikawa et al. (2013) investigated the daily activity patterns of the Mt. Dairoku

Table 6.2 A list of mammals at the University of Tokyo Hokkaido Forest

Order	Family	Genus	Species	Common name	References	Remarks
Soricomorpha	Soricidae	*Sorex*	*caecutiens*	Laxmann's shrew, Eurasian common shrew	Akiba et al. (2013)	
		Sorex	*unguiculatus*	Long-clawed shrew	Akiba et al. (2013)	
		Sorex	*gracillimus*	Slender shrew	Akiba et al. (2013)	
Chiroptera	Rhinolophidae	*Rhinolophus*	*nippon*	Greater horseshoe bat	Fukui (pers. com.)	
		Rhinolophus	*cornutus*	Japanese little horseshoe bat	Vertebrate Group, Biology Division, Fundamental Data Development Committee (2022)	
	Vespertilionidae	*Barbastella*	*pacifica*	Eastern barbastelle	Vertebrate Group, Biology Division, Fundamental Data Development Committee (2022)	
		Plecotus	*sacrimontis*	Japanese long-eared bat	Vertebrate Group, Biology Division, Fundamental Data Development Committee (2022)	
		Vespertilio	*sinensis*	Asian parti-colored bat	Vertebrate Group, Biology Division, Fundamental Data Development Committee (2022)	
		Myotis	*frater*	Fraternal Myotis	Vertebrate Group, Biology Division, Fundamental Data Development Committee (2022)	
		Myotis	*ikonnikovi*	Ikonnikov's Myotis	Vertebrate Group, Biology Division, Fundamental Data Development Committee (2022)	
		Myotis	*macrodactylus*	Japanese large-footed bat	Vertebrate Group, Biology Division, Fundamental Data Development Committee (2022)	
		Myotis	*petax*	Eastern water bat, Daubenton's bat	Vertebrate Group, Biology Division, Fundamental Data Development Committee (2022)	
		Murina	*hilgendorfi*	Hilgendorf's tube-nosed bat	Vertebrate Group, Biology Division, Fundamental Data Development Committee (2022)	
		Murina	*ussuriensis*	Ussurian tube-nosed bat	Vertebrate Group, Biology Division, Fundamental Data Development Committee (2022)	
Rodentia	Muridae	*Craseomys*	*rufocanus*	Gray red-backed vole	Akiba et al. (2013) and Shibano and Takeuchi (1983)	

		Craseomys	*rex*	Dark red-backed vole	Shibano and Takeuchi (1983)	
		Craseomys	*rutilus*	Northern red-backed vole	Akiba et al. (2013)	
		Apodemus	*speciosus*	Large Japanese field mouse	Akiba et al. (2013) and Shibano and Takeuchi (1983)	
		Apodemus	*argenteus*	Small Japanese field mouse	Akiba et al. (2013) and Shibano and Takeuchi (1983)	
		Rattus	*norvegicus*	Norway rat	Rokugo 50th Anniversary Commemorative Publication Editing Committee (1971) and Higashiyama Local History Editing Committee (1989)	Introduced
		Mus	*musculus*	House mouse	Rokugo 50th Anniversary Commemorative Publication Editing Committee (1971) and Higashiyama Local History Editing Committee (1989)	Introduced
	Sciuridae	*Sciurus*	*vulgaris*	Eurasian red squirrel	Oikawa and Matsui (2017) and Oikawa et al. (2022a, b)	
		Tamias	*sibiricus*	Siberian chipmunk	Oikawa et al. (2022a, b)	
		Pteromys	*volans*	Siberian flying squirrel, Russian flying squirrel	Oikawa and Matsui (2017)and Kadoya et al. (2010)	
Lagomorpha	Ochotonidae	*Ochotona*	*hyperborea*	Northern pika	Oikawa et al. (2013)	
	Leporidae	*Lepus*	*timidus*	Mountain hare	Oikawa and Matsui (2017) and Oikawa et al. (2022a)	
Carnivora	Canidae	*Vulpes*	*vulpes*	Red fox	Oikawa and Matsui (2017) and Oikawa et al. (2022a, b)	
		Nyctereutes	*procyonoides*	Raccoon dog	Oikawa and Matsui (2017) and Oikawa et al. (2022a, b)	

(continued)

Table 6.2 (continued)

Order	Family	Genus	Species	Common name	References	Remarks
		Canis	*familiaris*	Feral dog	Oikawa et al. (2022a)	Introduced
	Procyonidae	*Procyon*	*lotor*	Raccoon, North American raccoon	Oikawa and Matsui (2017) and Oikawa et al. (2022a, b)	Introduced
	Ursidae	*Ursus*	*arctos*	Brown bear	Oikawa and Matsui (2017) and Oikawa et al. (2022a, b)	
	Mustelidae	*Mustela*	*nivalis*	Least weasel	Rokugo 50th Anniversary Commemorative Publication Editing Committee (1971) and Higashiyama Local History Editing Committee (1989)	
		Mustela	*erminea*	Ermine, stoat	Rokugo 50th Anniversary Commemorative Publication Editing Committee (1971), Higashiyama Local History Editing Committee (1989) and Oikawa and Matsui (2009)	
		Neovison	*vison*	American mink	Oikawa et al. (2022a, b)	Introduced
		Martes	*zibellina*	Sable	Oikawa and Matsui (2017) and Oikawa et al. (2022a, b)	
	Felidae	*Felis*	*catus*	Feral cat	Oikawa and Matsui (2017) and Oikawa et al. (2022a, b)	Introduced
Artiodactyla	Cervidae	*Cervus*	*nippon*	Sika deer	Oikawa and Matsui (2017) and Oikawa et al. (2022a, b)	

Fig. 6.4 Ezo brown bear (*Ursus arctos yesoensis*). (Photo by Nozomi Oikawa, used with the UTHF's permission, licensed under CC BY-NC-ND 4.0)

Fig. 6.5 Ezo sika deer, *Cervus nippon yesoensis*. (Photo by Nozomi Oikawa, used with the UTHF's permission, licensed under CC BY-NC-ND 4.0)

population using camera traps and found that while there is variation among individuals and seasons, Japanese pika is more active at night.

Heim et al. (2021) utilized metabarcording techniques to analyze the diet of two sympatric bat species inhabiting the UTHF: Ikonnikov's Myotis, *Myotis ikonnikovi,* and the Ussurian tube-nosed bat, *Murina ussuriensis*. They found greater prey diversity in the diet of *M. ikonnikovi* compared to that of *M. ussuriensis*, suggesting that the former may be a more generalist predator. The diet of *M. ussuriensis* contains numerous Lepidopteran families, and the highly likelihood of prey items captured through gleaning in its diet, compared to *M. ikonnikovi,* indicates that *M. ussuriensis* may switch between the aerial-hawking and gleaning modes of foraging behavior.

Using camera trap data, Oikawa et al. (2022b) analyzed the diurnal activity patterns of medium and large mammals. The results revealed that sika deer and red foxes exhibit crepuscular activity, raccoon dogs are nocturnal, and brown bears display cathemeral behavior.

Extensive long-term surveys conducted by Oshida and colleagues on Siberian flying squirrels have revealed insights into habitat characteristics, nest behavior, and

Fig. 6.6 Northern pika, *Ochotona hyperborea yesoensis.* (Photo by Nozomi Oikawa, used with the UTHF's permission, licensed under CC BY-NC-ND 4.0)

Fig. 6.7 Japanese barbastelle, *Barbastella pacifica.* (Photo by Dai Fukui, Sep. 3, 2020)

microbial fauna in their gut (Kadoya et al. 2010; Marugame et al. 2010; Suzuki et al. 2011; Oshida et al. 2018; Liu et al. 2018, 2020; Kikuchi et al. 2023). Additionally, Kikuchi and Oshida (2023) uncovered arboreal nest-building behavior in the small Japanese field mouse (*Apodemus argenteus*), while Izumi et al. (2011) investigated habitat use by the Eurasian red squirrel (*Sciurus vulgaris orientis*).

In Hokkaido, a total of 53 species of terrestrial mammals, including 11 nonnative species, are distributed. Of these, 33 species have been recorded in the UTHF, suggesting that the area exhibits a high diversity of mammal species. This diversity is likely attributed to the environment variability resulting from altitudinal and geographical variation, as well as the presence of good habitat conditions under the Stand-based Silvicultural Management Systems.

6.4 Birds

Most studies and reports on birds in the University of Tokyo Hokkaido Forest were published after the 1970s (e.g., Arisawa 1976; Kojima and Arisawa 1983; Arisawa 1991). Notably, no comprehensive avifaunal surveys covering the entire area were conducted in the nineteenth century, with only one example (Arisawa and Shibano 1983) focusing on specific areas of the UTHF. In this century, the importance of accumulating fundamental data and conducting long-term monitoring has been recognized. Since 2004, the Bird Research Group of the University of Tokyo Forests has carried out line censuses and point censuses with the results published every 6 or 7 years. The first report (Bird Research Group of the University of Tokyo Forests 2009) recorded 13 orders, 37 families, 74 genera, and 110 species; the second report (Bird Research Group of the University of Tokyo Forests 2015) documented 15 orders, 40 families, 81 genera, and 115 species; and the most recent report (Bird Research Group of the University of Tokyo Forests (2022) recorded 15 orders, 40 families, 84 genera, and 121 species (Table 6.3 and Figs. 6.8 and 6.9). Among these are three species designated as the natural monuments of Japan: the black woodpecker (Fig. 6.10), Steller's sea eagle (Fig. 6.11), and white-tailed eagle.

Some insights have been gained into the ecology of black woodpecker. Arisawa (1991) summarized the nest site environment of black woodpecker from 33 years of observations, revealing Sakhalin fir as the most frequently used nesting tree, followed by Erman's birch and white birch. The nesting trees were characterized by straight trunks, large diameter, smooth bark, high branch-free heights, and slight inclinations. Furthermore, Kojima and Arisawa (1983) examined the black woodpecker's winter diet, revealing that it consists almost entirely of ants.

Table 6.3 A list of birds at the University of Tokyo Hokkaido Forest

Order	Family	Genus	Species (subspecies)	Common name	Remarks
Galliformes	Phasianidae	*Tetrastes*	*bonasia vicinitas*	Hazel grouse	
Anseriformes	Anatidae	*Cygnus*	*columbianus jankowskyi*	Tundra swan	
		Cygnus	*cygnus*	Whooper swan	
		Aix	*galericulata*	Mandarin duck	
		Anas	*platyrhynchos platyrhynchos*	Mallard	
		Anas	*zonorhyncha*	Eastern spot-billed duck	
		Anas	*crecca crecca*	Teal	
		Mergellus	*albellus*	Smew	
		Mergus	*merganser merganser*	Common merganser	
Columbiformes	Columbidae	*Streptopelia*	*orientalis orientalis*	Oriental turtle dove	
		Treron	*sieboldii sieboldii*	Japanese green-pigeon	
Pelecaniformes	Ardeidae	*Ardea*	*cinerea jouyi*	Grey heron	
Gruiformes	Rallidae	*Rallus*	*aquaticus indicus*	Water rail	
		Porzana	*fusca erythrothorax*	Ruddy-breasted crake	
		Gallinula	*chloropus chloropus*	Common moorhen	
Cuculiformes	Cuculidae	*Hierococcyx*	*hyperythrus*	Rufous hawk-cuckoo	
		Cuculus	*optatus*	Oriental cuckoo	
		Cuculus	*canorus telephonus*	Common cuckoo	
Caprimulgiformes	Caprimulgidae	*Caprimulgus*	*indicus jotaka*	Jungle nightjar	
Apodiformes	Apodidae	*Hirundapus*	*caudacutus caudacutus*	White-throated needltailed swift	
		Apus	*pacificus pacificus*	Pacific swift	
Charadriiformes	Charadriidae	*Charadrius*	*dubius curonicus*	Little ringed plover	
	Scolopacidae	*Scolopax*	*rusticola*	Eurasian woodcock	
		Gallinago	*solitaria japonica*	Solitary snipe	
		Gallinago	*hardwickii*	Latham's snipe	

		Actitis	*hypoleucos*	Common sandpiper	
Accipitriformes	Pandionidae	*Pandion*	*haliaetus haliaetus*	Western osprey	
	Accipitridae	*Milvus*	*migrans lineatus*	Black kite	
		Haliaeetus	*albicilla albicilla*	White-tailed eagle	Natural monument
		Haliaeetus	*pelagicus*	Steller's sea eagle	Natural monument
		Accipiter	*gularis gularis*	Japanese sparrowhawk	
		Accipiter	*nisus nisosimilis*	Eurasian sparrowhawk	
		Accipiter	*gentilis fujiyamae*	Northern goshawk	
		Buteo	*buteo japonicus*	Common buzzard	
		Buteo	*lagopus menzbieri*	Rough-legged buzzard	
		Nisaetus	*nipalensis orientalis*	Mountain hawk-eagle	
Strigiformes	Strigidae	*Otus*	*sunia japonicus*	Oriental scops owl	
		Strix	*uralensis japonica*	Ural owl	
		Ninox	*scutulata japonica*	Brown hawk-owl	
Coraciiformes	Alcedinidae	*Halcyon*	*coromanda major*	Ruddy kingfisher	
		Alcedo	*atthis bengalensis*	Common kingfisher	
		Megaceryle	*lugubris pallida*	Crested kingfisher	
Piciformes	Picidae	*Dendrocopos*	*kizuki seebohmi*	Japanese pygmy woodpecker	
		Dendrocopos	*minor amurensis*	Lesser spotted woodpecker	
		Dendrocopos	*leucotos subcirris*	White-backed woodpecker	
		Dendrocopos	*major japonicus*	Great spotted woodpecker	
		Dryocopus	*martius martius*	Black woodpecker	Natural monument
		Picus	*canus jessoensis*	Grey-headed woodpecker	
Falconiformes	Falconidae	*Falco*	*subbuteo subbuteo*	Eurasian hobby	
		Falco	*peregrinus japonensis*	Peregrine falcon	
Passeriformes	Laniidae	*Lanius*	*bucephalus bucephalus*	Bull-headed shrike	

(continued)

Table 6.3 (continued)

Order	Family	Genus	Species (subspecies)	Common name	Remarks
		Lanius	*excubitor bianchii*	Great grey shrike	
	Corvidae	*Garrulus*	*glandarius brandtii*	Eurasian jay	
		Nucifraga	*caryocatactes japonica*	Spotted nutcracker	
		Corvus	*corone orientalis*	Carrion crow	
		Corvus	*macrorhynchos japonensis*	Large-billed crow	
	Regulidae	*Regulus*	*regulus japonensis*	Goldcrest	
	Paridae	*Poecile*	*palustris hensoni*	Marsh tit	
		Poecile	*montanus restricus*	Willow tit	
		Poecile	*varius varius*	Varied tit	
		Periparus	*ater insularis*	Coal tit	
		Parus	*minor minor*	Japanese tit	
	Alaudidae	*Alauda*	*arvensis japonica*	Eurasian skylark	
	Hirundinidae	*Delichon*	*dasypus dasypus*	Asian house martin	
	Pycnonotidae	*Hypsipetes*	*amaurotis amaurotis*	Brown-eared bulbul	
	Cettiidae	*Cettia*	*diphone cantans*	Japanese bush warbler	
		Urosphena	*squameiceps*	Asian stubtail	
	Aegithalidae	*Aegithalos*	*caudatus japonicus*	Long-tailed tit	
	Phylloscopidae	*Phylloscopus*	*borealoides*	Sakhalin leaf-warbler	
		Phylloscopus	*coronatus*	Eastern crowned-warbler	
	Zosteropidae	*Zosterops*	*japonicus japonicus*	Japanese white-eye	
	Locustellidae	*Locustella*	*fasciolata amnicola*	Gray's grasshopper-warbler	
	Bombycillidae	*Bombycilla*	*garrulus centralasiae*	Bohemian waxwing	
		Bombycilla	*japonica*	Japanese waxwing	
	Sittidae	*Sitta*	*europaea asiatica*	Eurasian nuthatch	
	Certhiidae	*Certhia*	*familiaris daurica*	Treecreeper	

	Troglodytidae	*Troglodytes*	*troglodytes fumigatus*	Eurasian wren	
	Sturnidae	*Spodiopsar*	*cineraceus*	White-cheeked starling	
		Agropsar	*philippensis*	Chestnut-cheeked starling	
	Cinclidae	*Cinclus*	*pallasii pallasii*	Brown dipper	
	Muscicapidae	*Zoothera*	*sibirica davisoni*	Siberian thrush	
		Zoothera	*dauma aurea*	Scaly thrush	
		Turdus	*cardis*	Japanese thrush	
		Turdus	*obscurus*	Eyebrowed thrush	
		Turdus	*pallidus*	Pale thrush	
		Turdus	*chrysolaus chrysolaus*	Brown-headed thrush	
		Turdus	*naumanni eunomus*	Naumann's thrush	
		Luscinia	*akahige akahige*	Japanese robin	
		Luscinia	*calliope*	Siberian rubythroat	
		Luscinia	*cyane bochaiensis*	Siberian blue robin	
		Tarsiger	*cyanurus cyanurus*	Red-flanked bluetail	
		Saxicola	*torquatus stejnegeri*	African stonechat	
		Monticola	*solitarius philippensis*	Blue rock thrush	
		Muscicapa	*sibirica sibirica*	Dark-sided flycatcher	
		Muscicapa	*dauurica dauurica*	Asian brown flycatcher	
		Ficedula	*narcissina narcissina*	Narcissus flycatcher	
		Cyanoptila	*cyanomelana cyanomelana*	Blue-and-white flycatcher	
	Prunellidae	*Prunella*	*rubida*	Japanese accentor	
	Passeridae	*Passer*	*rutilans rutilans*	Russet sparrow	
		Passer	*montanus saturatus*	Eurasian tree sparrow	
	Motacillidae	*Motacilla*	*cinerea cinerea*	Grey wagtail	
		Motacilla	*alba lugens*	White wagtail	

(continued)

Table 6.3 (continued)

Order	Family	Genus	Species (subspecies)	Common name	Remarks
		Motacilla	*grandis*	Japanese wagtail	
		Anthus	*hodgsoni hodgsoni*	Olive-backed pipit	
	Fringillidae	*Fringilla*	*montifringilla*	Brambling	
		Chloris	*sinica minor*	Oriental greenfinch	
		Carduelis	*spinus*	Eurasian siskin	
		Carduelis	*flammea flammea*	Common redpoll	
		Carduelis	*hornemanni*	Arctic redpoll	
		Uragus	*sibiricus sanguinolentus*	Long-tailed rosefinch	
		Pinicola	*enucleator sakhalinensis*	Pine grosbeak	
		Loxia	*curvirostra japonica*	Common crossbill	
		Pyrrhula	*pyrrhula rosacea*	Eurasian bullfinch	
		Pyrrhula	*pyrrhula griseiventris*	Eurasian bullfinch	
		Coccothraustes	*coccothraustes japonicus*	Hawfinch	
		Eophona	*personata personata*	Japanese grosbeak	
	Emberizidae	*Emberiza*	*cioides ciopsis*	Meadow bunting	
		Emberiza	*rustica latifascia*	Rustic bunting	
		Emberiza	*elegans elegans*	Yellow-throated bunting	
		Emberiza	*spodocephala personata*	Black-faced bunting	
		Emberiza	*variabilis*	Grey bunting	

Cited from Bird research group of the University of Tokyo Forests (2022). English common name is referred from the Ornithological Society of Japan (2012)

Fig. 6.8 Hazel grouse, *Tetrastes bonasia.* (Photo by Nozomi Oikawa, used with the UTHF's permission, licensed under CC BY-NC-ND 4.0)

Fig. 6.9 Crested kingfisher, *Megaceryle lugubris.* (Photo by Nozomi Oikawa, used with the UTHF's permission, licensed under CC BY-NC-ND 4.0)

Fig. 6.10 Black woodpecker, *Dryocopus martius*. (Photo by Tetsuyuki Kobayashi, used with the UTHF's permission, licensed under CC BY-NC-ND 4.0)

Fig. 6.11 Steller's sea eagle, *Haliaeetus pelagicus*. (Photo by Nozomi Oikawa, used with the UTHF's permission, licensed under CC BY-NC-ND 4.0)

Table 6.4 A list of fish at the University of Tokyo Hokkaido Forest

Order	Family	Genus	Species	Common name	Remarks
Salmoniformes	Salmonidae	*Salvelinus*	*leucomaenis leucomaenis*	White-spotted char	
		Salvelinus		Brook trout	Introduced
		Salvelinus	*leucomaenis leucomaenis* x *fontinalis*		Hybrid
		Oncorhyncus	*mykiss*	Rainbow trout	Introduced
		Salvelinus	*malma*	Dolly varden	
Cypriniformes	Cobitidae	*Noemacheilus*	*barbatulus toni*	Siberian stone loach	
	Cyprinidae	*Tribolodon*	*hakonensis*	Japanese dace	
Scorpaeniformes	Cottidae	*Cottus*	*nozawae*	Sculpin	
Petromyzontiformes	Petromyzontidae	*Lethenteron*	*reissneri*	Brook lamprey	

Cited from Fukui et al. (2016)

6.5 Fishes

Until recently, no research has been conducted on the fish fauna of the UTHF. However, during the 2010s, studies on fishes in the Sorachi River water system, including the UTHF, were initiated, resulting in the publication of species lists for fish inhabiting the UTHF (Fukui et al. 2016, 2018; Fukui and Koizumi 2020; Suzuki et al. 2021; Ishiyama et al. 2023). The UTHF, located in the upper reaches of the Sorachi River drainage system, is primarily divided into three tributary drainage systems: the Furebetsu, Nunobe, and Nishitappu drainage systems. According to Fukui et al. (2016), seven species and one hybrid species were confirmed in the Furebetsu River system, eight species and one hybrid species were confirmed in the Nunobe River system, and seven species were confirmed in the Nishitappu River system (Table 6.4 and Fig. 6.12). Although some of these species have only been confirmed outside the boundaries of the UTHF, but because they are in the same water system and close together, they are likely to be inhibited within the UTHF area due to their proximity and presence in the same tributary system, and therefore, they are included in the species list. Fukui et al. (2016) also reported the presence of two introduced species, namely, Brook trout (*Salvelinus fontinalis*) and rainbow trout (*Oncorhyncus mykiss*), in all of the three tributary systems. Notably, F1 hybrids and post-F1 hybrids between Brook trout and the native species (white spotted char, *Salvelinus leucomaenis leucomaenis*: Fig. 6.13) have been confirmed in the Furebetsu and Nunobe River systems, indicating the spread of genetic

Fig. 6.12 White spotted char, *Salvelinus leucomaenis leucomaenis*. White bar shows 1 cm. (Photo by Mitsukazu Mitsugi)

Fig. 6.13 Fl hybrids or post-Fl hybrids between Brook trout and white spotted char. White bar shows 1 cm. (Photo by Mitsukazu Mitsugi)

contamination in the population of native species. There are also small ponds within the UTHF, although their total area is limited. The fish fauna in such environments have not been studied so far, but a systematic survey is currently underway of the 10-year plan initiated in 2021. It is anticipated that these surveys will contribute to further understanding of the fish fauna in the near future.

References

Akiba Y, Katayama R, Okamoto M, Ohkawa A, Hasegawa H, Asakawa M (2013) Collection records of endo- and exoparasites obtained from small mammals in the University of Tokyo Hokkaido Forest, Furano, Hokkaido, Japan. Bull Bio-geogr Soc Jpn 68:129–133

Arisawa H (1965a) First record of *Pachyligia dolosa* (Lepidoptera: Geometridae) from Hokkaido. Jpn Heterocerists' J 38:324

Arisawa H (1965b) Interesting Heterocera (Lepidoptera) from Yamabe, Hokkaido. Jpn Heterocerists' J 40:355–359

Arisawa H (1976) Syouwa 50-nendo kankyou-cyou itaku cyousa, tokutei cyourui cyousa: Kumagera [Survey on Designated Birds and Other Species Commissioned by Environmental Agency, Fiscal Year 1975: *Dryocopus martius*] , 35–85. (in Japanese)

Arisawa H (1991) Breeding nest density and the characteristics of nest tree of black woodpecker, *Dryocopus martius martius*. Bull Univ Tokyo For 84:21–37. (in Japanese with English summary)

Arisawa H, Shibano S (1983) Cyourui [Birds]. In: Shizen Kouen Sougou Cyousa (Furano Ashibetu Douritu Shizen Kouen) Houkokusho: Hokkaido Dai 3-bu Doubutsu [Comprehensive Survey of Natural Parks (Furano-Ashibetsu Prefectural Natural Park) Report: Hokkaido, Part 3: Fauna, Hokkaido Prefecture, 144–155. (in Japanese)

Biodiversity Center of Japan (2009) Summary report of Forest and Grassland Survey on Monitoring Sites 1000 Project in the 1st Term, Biodiversity Center of Japan, Ministry of the Environment, Government of Japan, Fujiyoshida (in Japanese with English summary).

Bird research group of the University of Tokyo Forests (2009) Bird list of university forests, the University of Tokyo. Misc Inf Univ Tokyo For 48:103–131. (in Japanese)

Bird research group of the University of Tokyo Forests (2015) Bird list of university forests, the University of Tokyo 2nd edition (August, 2008 to March, 2014). Miscellaneous Information of the University of Tokyo Forests 57:75–102. (in Japanese)

Bird research group of the University of Tokyo Forests (2022) Bird list of university forests, the University of Tokyo 3rd edition (April, 2008 to March, 2021). Misc Inf Univ Tokyo For 57:75–102. (in Japanese)

Fukui S, Koizumi I (2020) Hybrids as potential mediators spreading non-native genes: comparison of survival, growth, and movement among native, introduced and their hybrid salmonids. Ecol Freshw Fish 29:280–288

Fukui S, May-McNally SL, Katahira H, Kitano S, Koizumi I (2016) Temporal change in the distribution and composition of native, introduced, and hybrid charrs in northern Japan. Hydrobiologica 783:309–316

Fukui S, May-McNally SL, Taylor EB, Koizumi I (2018) Maladaptive secondary sexual characteristics reduce the reproductive success of hybrids between native and non-native salmonids. Ecol Evol 8:12173–12182

Heim O, Puisto A, Sääksjärvi I, Fukui D, Vesterinen EJ (2021) Dietary analysis reveals differences in the prey use of two sympatric bat species. Ecol Evol 11:18651–18661

Higashiyama Local History Editing Committee (ed) (1989) Higashiyama Kyoudoshi [Higashiyama Local History], Higashiyama 80th Anniversary Commemorative Project Executive Committee, Japan, 594 pp

Insect Group, Biology Section, Long-term Ecological Data Division, The University of Tokyo Forests (2024) A catalog of ground-dwelling beetles captured by pitfall traps in the University of Tokyo Forests (2008–2012). Misc Inf Univ Tokyo For 68:XX

Ishiyama N, Sueyoshi M, Molinos JG, Iwasaki K, Negishi JN, Koizumi I, Nagayama S, Nagasaka A, Nagasaka Y, Nakamura F (2023) Underlying geology and climate interactively shape climate change refugia in mountain streams. Ecol Monogr 93:e1566

Izumi I, Matsui M, Okahira T, Hayashi Y, Oshida T (2011) Preliminary survey of habitat use by *Sciurus vulgaris orientis* in a Natural Forest of Hokkaido Island, Japan. Mammal Study 36:109–112

Kadoya N, Iguchi K, Matsui M, Okahira T, Kato A, Oshida T, Hayashi Y (2010) A preliminary survey on nest cavity use by Siberian flying squirrels, *Pteromys volans orii*, in forests of Hokkaido Island, Japan. Russ J Theriol 9:27–32

Kamata K, Iguchi K, Sanguansub S, Maneerat T (2014) Infestation by insect borers on Betula maximowicziana following successive years of severe defoliation by *Caligula japonica* (Lepidoptera: Saturniidae) with special reference to *Xyleborus seriatus* Blandford (Coleoptera: Curculionidae: Scolytinae). For Pests (Japan) 72:42–48

Kikuchi H, Oshida T (2023) A case of arboreal nest building in the small Japanese field mouse (*Apodemus argenteus*). Mammalia 87:214–217

Kikuchi H, Akasaka T, Asari Y, Oshida T (2023) Communal nesting behaviour of Siberian flying squirrels during the non-winter season. Ethology 129:499–506

Kojima K, Arisawa H (1983) Habitat and food habit of the black woodpecker *Drycocopus martius* in Hokkaido. Tori 32:109–111

Liu P-Y, Cheng A-C, Huang S-W, Chang H-W, Oshida T, Yu H-T (2018) Variations in gut microbiota of Siberian flying squirrels correspond to seasonal phenological changes in their Hokkaido subarctic forest ecosystem. Microb Ecol 78:223–231

Liu P-Y, Cheng A-C, Huang S-W, Lu H-P, Oshida T, Liu W, Yu H-T (2020) Body-size scaling is related to gut microbial diversity, metabolism and dietary niche of arboreal folivorous flying squirrels. Sci Rep 10:7809

Marugame M, Izumi I, Matsui M, Okahira T, Oshida T, Hayashi Y (2010) Cavity resources for Siberian flying squirrel, *Pteromys volans orii*, in two different habitats in Hokkaido, Japan. Russ J Theriol 9(1):39–43

Nishiguchi C (1957a) On bark beetles attacking pines. Misc Inf Univ Tokyo For 12:69–72

Nishiguchi C (1957b) On bark beetles attacking Douglas fir. Misc Inf Univ Tokyo For 12:69–72

Nishiguchi C (1958) On the communities of Populus defoliators and their seasonal change. Bull Univ Tokyo For 54:191–198

Nishiguchi C (1959) Species compositions of swarms on spruce bark-and wood-inhabiting beetles and their seasonal change. J Jpn For Soc 41:270–274

Nishiguchi C (1960a) Insects pests found on Populus spp at arboretum of the University of Tokyo Hokkaido Forest I: Macrolepidoptera. POPURA 7:10–11

Nishiguchi C (1960b) Insects pests found on Populus spp at arboretum of the University of Tokyo Hokkaido Forest II: leaf beetles. POPURA 8:5–6

Nishiguchi C (1960c) Synökologische Beobachtungen über die Befall der stehenden Fichte durch Borkenkäfer. J Jpn For Soc 42:64–73

Nishiguchi C (1960d) Die Verteilungsformen der Rindenbrutern an den Kiefern im Hokkaido Forst der Universitat Tokyo. Misc Inf Univ Tokyo For 13:27–38

Nishiguchi C (1962a) Insects pests found on Populus spp at arboretum of the University of Tokyo Hokkaido Forest III: Microlepidoptera. POPURA 14:9

Nishiguchi C (1962b) Insects pests found on Populus spp at arboretum of the University of Tokyo Hokkaido Forest IV: wood boring Lepidoptera, Aegeria asamaensis Hampson (Lepidoptera: Sesiidae) Microlepidoptera. POPURA 15:2–4

Oikawa N, Matsui M (2009) Tokyo Daigaku Hokkaido Ensyurin ni okeru Ezo-nakiusagi no Seisoku Joukyou [habitat status of the Northern Pika in the University of Tokyo Hokkaido Forest] Northern Foresty. Japan 61:11–14

Oikawa N, Matsui M (2017) Monitoring report of vertebrates in the University of Tokyo Hokkaido Forest, using camera traps (April, 2011–March, 2016). Misc Inf Univ Tokyo For 59:85–92. (in Japanese)

Oikawa N, Matsui M, Hirakawa H (2013) Notably high nighttime activity in the autumn—daily rhythm of activity in the northern pika (*Ochotona hyperborea*) in a stunted subalpine forest in Central Hokkaido, Japan. Mamm Sci (Honyurui Kagaku) 53:79–87. (in Japanese with English abstract)

Oikawa N, Kobayashi T, Katahira K (2022a) Monitoring report of the vertebrates using camera traps in the University of Tokyo Hokkaido Forest (April, 2016–March, 2021). Misc Inf Univ Tokyo For 65:21–23. (in Japanese)

Oikawa N, Matsui M, Fukui D, Hirakawa H (2022b) Results of the mammalian faunal survey in Central Hokkaido using camera traps. Boreal For Res 70:65–68

Oshida T, Komoto A, Shibatani M, Yoshikawa Y, Sato D (2018) Do Siberian flying squirrels reuse nest materials made by other individuals? Acta Zool Acad Sci Hung 64:185–192

Peng Y (2022) Bark and ambrosia beetle community in Hokkaido, Japan: temporal dynamics of flying adults, succession, and relationship with host plant phylogeny. Ph. D dissertation,. The University of Tokyo

Peng Y, Buranapanichpan A, Kamata N (2022) Succession of ambrosia beetles colonizing the logs of fallen alder and birch trees. Insects 13:223

Rokugo 50th Anniversary Commemorative Publication Editing Committee (ed) (1971) Rokugo No 50 Nen [Rokugo 50-year history], Rokugo 50th anniversary cooperation, Japan, 291 pp

Shibano S, Takeuchi M (1983) Nezumi-rui [Rodents] In: Shizen Kouen Sougou Cyousa (Furano Ashibetu Douritu Shizen Kouen) Houkokusho: Hokkaido Dai 3-bu Doubutsu [comprehensive survey of natural parks (Furano-Ashibetsu Prefectural Natural Park) report: Hokkaido, part 3: Fauna, Hokkaido Prefecture, 174–181

Suzuki M, Kato A, Matsui M, Okahira T, Iguchi K, Hayashi Y, Oshida T (2011) Preliminary estimation of population density of the Siberian flying squirrel (*Pteromys volans orii*) in natural forest of Hokkaido, Japan. Mammal Study 36:155–158

Suzuki K, Ishiyama N, Koizumi I, Nakamura F (2021) Combined effects of summer water temperature and current velocity on the distribution of a cold-water-adapted sculpin (*Cottus nozawae*). Water 13:975

Takahashi Y, Inukai M, Iguchi K, Takahashi I, Yamamoto H (1997) A case of forest damage caused by sika deer (*Cervus nippon yesoensis*, HEUDE) in Iwanazawa experimental plot of the Tokyo university Forest in Hokkaido. Trans Meet Hokkaido Br Jpn For Soc 45:84–87

The Ornithological Society of Japan (2012) Check-list of Japanese birds, 7th revised edition. pp 438

Toki W, Ishihama N, Monji S, Matsuoka F, Yuzawa N, Iguchi K, Suzuki S (2024) Fauna of flower longicorn beetles (Coleoptera, Cerambycidae) in the University of Tokyo Hokkaido Forest. Misc Inf Univ Tokyo For 70:13–23

Umeya K (2003) How to rear insects. Japan Association for Techno-innovation in agriculture, forestry and fisheries, https://www.jataff.or.jp/konchu/breeding/1_1.html#top. Last visited on OCT 07, 2024

Vertebrate Group, Biology Division, Fundamental Data Development Committee (2022) Faunal survey report of bats (Mammalia: Chiroptera) in the University of Tokyo Forests between 2013–2020. Misc Inf Univ Tokyo For 65:111–122. (in Japanese)

Wallon S, Rigal F, Melo CD, Elias RB, Borges PAV (2024) Unveiling arthropod responses to climate change: a functional trait analysis in intensive pastures. Insects 15:677

Part III
Forest Management and Silviculture

Chapter 7
The Stand-Based Silvicultural Management System: An Introduction

Toshiaki Owari

Abstract This chapter provides an introduction to the Stand-based Silvicultural Management System (SSMS), a system that has been implemented by the University of Tokyo Hokkaido Forest (UTHF) since 1958. Under the SSMS, forests are categorized into distinct forest types based on tree density, species composition, diameter distribution, and the natural regeneration status of trees. Silvicultural operations are tailored to the specific conditions of each forest type. The SSMS is an integrated method that initially drew upon existing experiences and knowledge of natural forest management from both Japan and abroad. The primary goal of the SSMS is to harmonize and enhance dual forest functions: a public function in environmental conservation and an economic function in the production of forest products through proper management practices. The SSMS addresses natural forests with various types based on six fundamental principles: (1) facilitating the transition to and sustaining a preclimax phase for each stand, (2) avoiding intensive forest operations, (3) adapting to the structural and site conditions of each stand, (4) maintaining high-storied and multistoried stands, (5) retaining trees with superior traits, and (6) nurturing mixed conifer-broadleaf forests.

Keywords Conceptual development · Mixed conifer-broadleaf forest · Natural forest management · Silvicultural principles · The University of Tokyo Hokkaido Forest

T. Owari (✉)
The University of Tokyo Hokkaido Forest, Graduate School of Agricultural and Life Sciences, The University of Tokyo, Furano, Japan
e-mail: owari@g.ecc.u-tokyo.ac.jp

T. Owari et al. (eds.), *Integrative Forest Management and Silviculture*,
Ecological Research Monographs, https://doi.org/10.1007/978-981-95-1017-7_7

7.1 What Is the Stand-Based Silvicultural Management System?

The University of Tokyo Hokkaido Forest (UTHF) has been utilizing a distinctive approach known as the "Stand-based Silvicultural Management System" (SSMS, or "Rinbun Segyo Ho" in Japanese) for over 65 years (Fig. 7.1). The SSMS involves subdividing the forest into distinct segments, each characterized by its tree density, species composition, diameter distribution, and the natural regeneration status of the trees. Silvicultural operations, including harvesting, planting, and tending, are then conducted according to the specific conditions of each segment (i.e., forest type) (refer to Chap. 8 for details). This method is highly regarded, both in Japan and internationally, as an approach to sustainable forest management and planning that effectively balances the use of renewable natural resources with the conservation of ecological functions (Watanabe and Sasaki 1994). Nobukiyo Takahashi (Fig. 7.2), who became a professor in 1954 and served as the 5th Director of the UTHF from 1942 to 1974, initiated the SSMS. He received the Duke of Edinburgh Prize of the Japan Academy in 1992 for his outstanding research and contributions to natural forest management and natural environment preservation (The Japan Academy). In 2014, the UTHF was designated as a "Forestry Heritage" by the Japanese Forest Society, recognizing and preserving the long-term and large-scale practices of natural forest management in Japan for future generations (The Japanese Forest Society).

Fig. 7.1 The mixed conifer-broadleaf forest managed by the Stand-based Silvicultural Management System. (Photo by the UTHF, licensed under CC BY-NC-ND 4.0)

Fig. 7.2 Nobukiyo Takahashi (1914–2002). (Photo by the UTHF, licensed under CC BY-NC-ND 4.0)

7.2 The Roots of the Stand-Based Silvicultural Management System

Takahashi conceived the SSMS with its fundamental principles and silvicultural guidelines (see Sect. 7.3) in 1955 (Takahashi 1961). The following year, in 1956, he implemented the SSMS at the Nunobe Silvicultural Experiment Site within the UTHF and confirmed its viability as a silvicultural management method for natural forests (Takahashi 2001). Subsequently, in 1958, the SSMS was formally incorporated into the 8th management plan (1958–1967) of the UTHF. This marked the initiation of a business-scale implementation of the SSMS across a vast forest area spanning approximately 20,000 hectares. Prof. Takahashi conceived the SSMS as a synthesis of the following three knowledge and experiences (Takahashi 1961).

7.2.1 Trial and Error from Previous Silvicultural Experiments

After beginning his work at the UTHF in 1938, Takahashi conducted silvicultural experiments in natural forests under the guidance of his former teachers, Prof. Masao Yoshida, Prof. Kentaro Nakamura, and Prof. Ichizo Mine (Takahashi 2001). In 1940, along with his colleague Miichiro Iwamoto, he initiated experimental research on silvicultural treatments in Compartment 108 on Jinja-yama within the UTHF (Fig. 7.3) (Takahashi 1961, 2001). When Takahashi was a student, Prof. Nakamura, who had studied German forestry at the University of Freiburg and later returned to Japan, gave him a lecture on the concept of *Dauerwald* (management form that requires continuous cover of trees at all time; Möller 1922; Puettmann et al. 2009; Yonemura 1984). Inspired by German forestry practices, he embarked on partial cutting trials to nurture natural forests. However, each attempt ended in failure to achieve natural regeneration after cutting due to overgrowth of *Sasa* dwarf bamboos (refer to Chap. 5), leading to a prolonged period of trial and error (Takahashi 2003).

It was not until more than a decade later, in fall 1953, that Takahashi came up with the idea that would later lead to the SSMS. At that time, he was walking in the forest. He recalled hearing the voice of a large Yezo spruce tree and learning the wisdom of "working on each small forest area" and "considering the needs of wild animals" (Yonemura 1984). He then realized that natural forests are composed of various types of stands and that silvicultural operations need to be tailored to the specific conditions of each stand type (Takahashi 2003).

7.2.2 Silvicultural Precedents in the Former Imperial Forest and the Hokkaido Prefectural Forest

Takahashi conceived the silvicultural management method by drawing insights from publications and empirical knowledge from the 1920s and 1930s, specifically from the work of engineers at the former Imperial Forest and the Hokkaido Prefectural Forest. Both organizations were pioneers in natural forest management compared to other forest management organizations in Hokkaido, northern Japan (Takahashi 1961; Shibata 1988). It appears that he drew inspiration from the stand-based silvicultural planning work in the former Imperial Forest (including the classification into tending, regeneration, and improvement stands), the practice of planting for assisted regeneration in the former Imperial Forest and the Hokkaido Prefectural Forest, and the tree marking method for selection cutting by Shoichi Tsumura (1926), a forest engineer of the Hokkaido Prefectural Government (Fig. 7.4) (Ohgane 1981; Takahashi 1961).

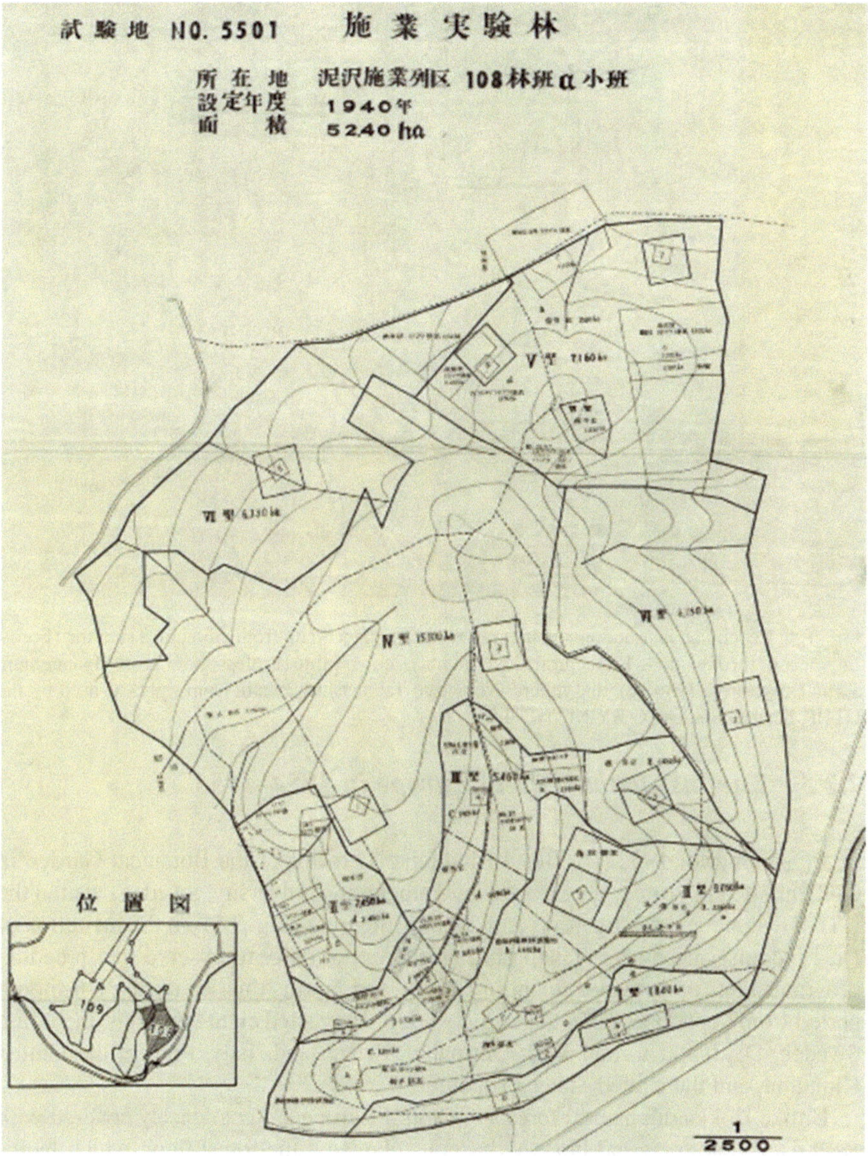

Fig. 7.3 Map of the Jinja-yama Silvicultural Experiment Forest (Compartment 108) that was established in 1940. The area was divided by seven segments according to forest type, and silvicultural treatments with measurement plots were designated for each segment. (Source: The University of Tokyo Hokkaido Forest)

Fig. 7.4 Receiving instruction on natural forest management from Shoichi Tsumura (Forest Engineer of the Hokkaido Prefectural Government), Ichizo Mine (Professor of Forest Management at the University of Tokyo), and others. Nobukiyo Takahashi (seventh from right). (Photo by the UTHF, licensed under CC BY-NC-ND 4.0)

7.2.3 The Inspection Trip to Europe in 1954–1955

In the summer of 1952, Dr. Bertil Lindquist (Director of the Botanical Garden in Gothenburg and Professor of Botany at Gothenburg University, Sweden) visited the UTHF to conduct investigations and collect elite trees (Fig. 7.5). At the invitation of Dr. Lindquist and others, Takahashi had the opportunity to observe tree breeding programs and forest management in Europe and North America during a 6-month period from September 1954 to February 1955. He visited eight countries, including Sweden, Denmark, former West Germany, Switzerland, Italy, France, the United Kingdom, and the United States (Yonemura 1984).

Unlike Hokkaido, natural forests in central Europe are not densely covered with *Sasa* dwarf bamboos and large herbaceous plants on the forest floor, which facilitated natural regeneration (Fig. 7.6; refer to Chap. 5). Takahashi recognized the challenges of managing natural forests in Hokkaido and realized that directly applying silvicultural methods from Europe, where trees naturally regenerate easily after cutting, would not be effective in promoting natural regeneration in Hokkaido (Takahashi 2000; Yonemura 1984). During his trip to Switzerland, he met Prof. Hans Leibundgut at Swiss Federal Institute of Technology (ETH) Zurich, who told him that the mean annual increment of the experimental forest managed by selection system was 16 m^3 per hectare. The fact convinced him that high stand growth

Fig. 7.5 Dr. Bertil Lindquist's visit to the UTHF in 1952. Lindquist (second from right), Taizo Inokuma (Professor of Forest Botany at the University of Tokyo, third from right), and Nobukiyo Takahashi (fifth from right). (Photo by the UTHF, licensed under CC BY-NC-ND 4.0)

Fig. 7.6 A forest in central Europe (Couvet, Switzerland) where natural regeneration is facilitated. (Photo by Toshiaki Owari)

could be achieved even in natural forests (Takahashi 1961). Plantation forests in former West Germany, which were subjected to repeated clear-cutting over large areas, experienced reduced growth, disease and insect pests, and frequent wind damage. After learning about the detrimental effects of extensive clear-cutting and simultaneous plantation establishment, Prof. Takahashi determined that forestry in Hokkaido should avoid making the same mistake (Takahashi 2001).

As mentioned above, Takahashi drew heavily from previous examples of natural forest management practices when conceiving the SSMS (Watanabe 2003). The SSMS is not entirely original; it represents a technique for natural forest management that has been refined through the incorporation of existing experiences and knowledge from both Japan and abroad (Miyamoto 2002; Takahashi 1961).

7.3 Fundamental Principles of the Stand-Based Silvicultural Management System

Forests serve a dual purpose, with a public function in environmental conservation and an economic function in producing timber and other forest products. The effectiveness of each of these functions relies on the structure of the forest as well as human interactions with it. Proper forest management is essential to enhance both functions for the future. The SSMS is built on this concept, addressing natural forests comprised of various forest types based on the following six fundamental principles (Takahashi 2001):

7.3.1 Principle 1: Facilitating the Transition to and Sustaining a Preclimax Phase for Each Stand

The forest management practices aim to expedite each component of the natural forest to reach the preclimax phase promptly. Forests in the intermediate phase are managed to progress toward the preclimax phase, while those in the preclimax phase are managed to maintain that state for an extended period.

The term "climax" refers to a dynamic state of equilibrium reached in the final stage of community succession. In a climax forest (Fig. 7.7a), the increase in biomass accumulation from tree growth is nearly offset by the loss due to tree mortality, primarily caused by the high number of old, overmature trees. As a result, by harvesting these aging trees with declining growth, the upper layer of the forest canopy is rejuvenated with younger, more vigorous trees, leading to growth rates that are expected to exceed mortality. In a vigorously growing forest, which is one

Fig. 7.7 (**a**) A forest stand in the climax phase (left) and (**b**) in the preclimax phase (right). (Photo by Toshiaki Owari)

step removed from climax phase and is referred to as the "preclimax" phase[1] (Fig. 7.7b), harvesting continues to maintain this phase for an extended period, with canopy trees being systematically replaced by new ones.

7.3.2 Principle 2: Avoiding Intensive Forest Operations

In natural forests, practices that cause severe and widespread destruction of ecosystems should be avoided as much as possible. Forest management practices should be designed to minimize the destruction and loss of forest ecosystems composed of living and nonliving organisms (Fig. 7.8). This is important to quickly restore the balance of the ecosystem (including small animals and microorganisms in the ground) that has been temporarily disturbed by the operation, from within and around the forest. It also has the effect of controlling the loss of water and nutrients in the forest area where the operation is carried out, as well as damage caused by weather (wind, snow, and cold), pathogens, insects, wildlife, and other factors.

[1] Note that this definition of "preclimax" differs from the original concept proposed by Clements (1916).

Fig. 7.8 A forest stand immediately after the selection cutting operation. (Photo by Toshiaki Owari)

7.3.3 Principle 3: Adapting to the Structural and Site Conditions of Each Stand

Natural forests are composed of a myriad of different forest segments. Depending on the structure and dynamics of each segment, appropriate practices should be implemented to further develop the overall function of the forest. If a forest is uniformly managed over a wide area by exclusively adopting a single textbook method, such as selection cutting, thinning, or clear-cutting, there will invariably be some inconsistencies and unnecessary treatments, hindering the forest's development. It is also advisable not to be overly rigid about maintaining a specific stand structure. Each forest stand (Fig. 7.9) has its natural direction of development. The most effective way to manage a forest is to nurture it in harmony with nature, rather than working against it.

Fig. 7.9 Different forest types: (**a**) Sakhalin fir (upper left), (**b**) Yezo spruce (upper right), and (**c**) broad-leaf species (lower) are dominant. (Photo by Toshiaki Owari)

7.3.4 Principle 4: Maintaining Tall, Multistoried Stands

To guide natural forests toward becoming tall, multistoried forests with the highest integrated functions, especially the upper canopy layer that receives the initial sunlight, should be managed to achieve its maximum quantitative and qualitative production capacity. In general, the total function of forests, which combines both timber production and environmental conservation, is (1) greater in mature forests than in young ones, (2) higher in forests with an appropriate stand density compared to sparse forests, (3) elevated in mixed forests as opposed to single-species forests, and (4) enhanced in multilayered forests rather than single-layer forests. A multistoried forest (Fig. 7.10), being one step removed from the climax phase, is expected to exhibit the highest biomass production per unit area and can be considered an ideal state.

Fig. 7.10 A mixed forest with multiple canopy layers dominated by Sakhalin fir. (Photo by Toshiaki Owari)

7.3.5 Principle 5: Retaining Trees with Superior Traits

Trees with unfavorable stem traits possibly due to genetic factors are selectively harvested and removed from the forest (Fig. 7.11), while those with superior traits are retained and managed to enhance the overall development of the forest. The traits of individual trees within a forest can be determined by the combined effects of genes and environmental factors. Although it is often uncertain whether poor traits are due to genetics or environmental factors, some forest sites seem to have populations with inherently poor genetic traits. The silvicultural management of naturally regenerated forests can be perceived as the process of artificially selecting for desirable genetic traits within the tree population of the forest. In the case of broadleaf trees, considerations include the straightness of the trunk, branch angle and thickness, the presence or absence of disease, and the overall material quality of the trees.

Fig. 7.11 Tree marking for harvest involves prioritizing trees with unfavorable traits. (Photo by the UTHF, licensed under CC BY-NC-ND 4.0)

7.3.6 Principle 6: Nurturing Mixed Conifer-Broadleaf Forests

The objective is to preserve the land's productive capacity and foster the development of robust, mixed forests composed of both conifer and broadleaf trees, which exhibit high resistance to various forms of damage. Natural forests are inherently self-fertilizing, meaning they enhance the biological, physical, and chemical properties and soil fertility through long-term cycles of material recycling. Additionally, they are self-regulating, as complex ecosystems within these forests naturally inhibit significant growth of certain pathogens and insects. Resistance to diseases and insect outbreaks is greater in mixed forests than in single-species forests, and in multistoried forests than in single-storied ones. It's worth noting that older trees are less resistant to fungal and insect damage (Fig. 7.12). The primary aim of harvesting older trees is not only to increase the overall growth and economic value of the forest area but also to establish a perpetually healthy forest. Furthermore, broadleaf trees in natural forests play a protective role in preventing soil degradation and increasing resistance to wind damage, diseases, and insect damage. A healthy forest represents a state in which numerous biological communities are well-balanced, and material cycles within the ecosystem are functioning normally.

Fig. 7.12 Old Sakhalin fir tree with advanced fungal decay. (Photo by Toshiaki Owari)

7.4 The Spirit of the Stand-Based Silvicultural Management System

In 1958, when the UTHF commenced experimenting with the SSMS, the Forestry Agency initiated the National Forest Productivity Enhancement Plan, also known as the Forestry Power Enhancement Plan. To meet the domestic demand for timber, the government decided to clear extensive areas of natural forests and simultaneously reforest these areas, a practice referred to as "afforestation expansion." While many regions were promoting uniform and aggressive forestry practices, the UTHF's resolute efforts to create ideal forests, guided by a strong conviction and unaffected by prevailing trends of the time, have earned it the high reputation it enjoys today (Shibata 1988). Watanabe (2003) asserted that the primary reason for the success of the SSMS was its ability to "adapt outstanding analysis and proposals on natural forest management by Prof. Nakamura, one of his teachers, to the actual forest conditions in the Furano region, where the UTHF is located."

The spirit of the SSMS planted by Takahashi has taken firm and thick roots in each and every staff member of the UTHF, and even now, more than 65 years later, it continues to support the forest management of the UTHF. As with the trees that make up the natural forests, the bearers of silvicultural operations will be changing from generation to generation, but I hope that the spirit of the SSMS will be passed on to future generations, along with the rich forests of the UTHF.

Acknowledgments The author would like to thank current and former staff members at the UTHF for their continuous and great efforts in managing the forest under the SSMS. The description in this chapter is mainly based on Owari (2014), with textual improvements and photo replacements.

References

Clements FE (1916) Plant succession: an analysis of the development of vegetation. Carnegie Institution of Washington, Washington DC

Miyamoto Y (2002) A critical study of "Rinbunsegyoho". For Econ 55(9):12–17. https://doi.org/10.19013/rinrin.55.9_12. (in Japanese with English title)

Möller A (1922) Der Dauerwaldgedanke: Sein Sinn und seine Bedeutung. Springer, Berlin/Heidelberg. https://doi.org/10.1007/978-3-642-50866-0

Ohgane E (1981) Rinbun Segyo Ho [The Stand-based Silvicultural Management System]. In: Ohgane E (ed) Nihon no takubatsu [Selection system in Japan]. Nihon Ringyo Chosa-kai, Tokyo, pp 239–240. (in Japanese)

Owari T (2014) Hokkaido Enshurin no Rinbun Segyo Ho [The Stand-based Silvicultural Management System in the Hokkaido Forest]. In: The University of Tokyo Forests Press (ed) Jukai wo yuku: Furano, Tokyo Daigaku Enshurin no mori zukuri [Traveling through the sea of trees: forest management in the University of Tokyo Hokkaido Forest, Furano]. The University of Tokyo Forests Press, Tokyo, pp 48–65. (in Japanese)

Puettmann KJ, Coates KD, Messier C (2009) A critique of silviculture: managing for complexity. Island Press, Washington

Shibata S (1988) Studies on the management of natural forests: an experiment of natural forest management in the Tokyo University Forest in Hokkaido. Bull Tokyo Univ For 80:269–397. (in Japanese with English summary)

Takahashi K (2000) Mori to ikiru: Dorogame-san to Tokyo Daigaku Hokkaido Enshurin [Living with the forest: Mr. Dorogame and the University of Tokyo Hokkaido Forest]. Popular Publishing, Tokyo. 143 pp, (in Japanese)

Takahashi N (1961) Rinbun Segyo Ho gairon: Enshurin ni okeru rekishi to jikken [Introduction to the Stand-based Silvicultural Management System: history and experiments at the University Forest]. Hoppo Ringyo [Northern Forestry] 13:105–109. (in Japanese)

Takahashi N (2001) Rinbun Segyo Ho: sono kangae to jissen (kaitei ban) [The Stand-based Silvicultural Management System: its Theory and Practices (revised ed.)]. Log Bee, Sapporo. 125 pp, (in Japanese)

Takahashi N (2003) Dorogame-san no mori zukuri [Forest management by Dorogame-san]. In: Dorogame-san saigo no hanashi – yume wo hagukumu Furano no mori zukuri [The last story of Dorogame-san: Forest development in Furano to foster dreams]. Shin Shisaku Sya, Tokyo, pp 69–100. (in Japanese)

The Japan Academy. https://www.japan-acad.go.jp/en/activities/jyusho/081to090.html#anker002

The Japanese Forest Society. https://www.forestry.jp/forestryheritage/2014-3086/

Tsumura S (1926) Shinrin ni taisuru mikata oyobi atsukaikata [How to view and treat forests]. Hokkaido Ringyo-kai, Sapporo. 198 pp, (in Japanese)

Watanabe S (2003) Rinbun Segyo Ho no kako, genzai, mirai: ni-zu ni oji shinka shi tsuzukeru tennen-rin segyo no shiten kara mita gijutsu ron [The past, present, and future of the Stand-based Silvicultural Management System: a technical theory from the viewpoint of natural forest management practices continuously deepened in response to needs]. In: Dorogame-san saigo no hanashi – yume wo hagukumu Furano no mori zukuri [The last story of Dorogame-san: Forest development in Furano to foster dreams]. Shin Shisaku Sya, Tokyo, pp 123–148. (in Japanese)

Watanabe S, Sasaki S (1994) The silvicultural management system in temperate and boreal forests: a case history of the Hokkaido Tokyo University Forest. Can J For Res 24(6):1176–1185. https://doi.org/10.1139/x94-155

Yonemura K (1984) Mori no hito: Dorogame-san [The forester: Mr. Dorogame]. Shunjuysha Publishing, Tokyo. 193 pp (in Japanese)

Chapter 8
Implementing the Stand-Based Silvicultural Management System

Toshiaki Owari

Abstract This chapter presents the fundamental techniques and operational procedures for the Stand-based Silvicultural Management System (SSMS), as implemented by the University of Tokyo Hokkaido Forest. The SSMS categorizes forests into six types: (1) Selection Cutting Forest, (2) Young Growth Forest, (3) Sparse Forest, (4) Secondary Forest, (5) Artificially Regenerated Forest, and (6) Restricted Forest, further subdivided based on dominant tree species, disturbance history, and/or regeneration status. Under the SSMS, natural forests are primarily managed using the single-tree selection system. In Conifer-dominant Selection Cutting Forest, the maximum cutting rate, represented by the volume ratio of growing stock, is set at 16–17%, ensuring it aligns with growth rates. Silvicultural management planning in each treated area follows these steps: (1) forest type classification survey, (2) plot inventory, (3) in-field review meeting, (4) tree marking, (5) harvesting operation, and (6) harvesting supervision. During tree marking, the selection of trees for harvest is carried out to enhance forest health and promote the regeneration and growth of successional juvenile trees. Moderate harvesting practices under the SSMS are believed to improve stand growth by enhancing the light environment, thereby fostering tree regeneration and growth.

Keywords Forest type · Operational procedure · Stand development · Selection system · Tree marking

8.1 Introduction

The Stand-based Silvicultural Management System (SSMS) has been implemented by the University of Tokyo Hokkaido Forest (UTHF) since 1958, in accordance with its forest management plan. This system is currently integrated into the 14th

T. Owari (✉)
The University of Tokyo Hokkaido Forest, Graduate School of Agricultural and Life Sciences, The University of Tokyo, Furano, Japan
e-mail: owari@g.ecc.u-tokyo.ac.jp

T. Owari et al. (eds.), *Integrative Forest Management and Silviculture*, Ecological Research Monographs, https://doi.org/10.1007/978-981-95-1017-7_8

education and research plan (The University of Tokyo Forests 2022; see Chap. 2). The SSMS is underpinned by two principal planning methodologies. The first is a spatially explicit forest type classification, which segments the forest into distinct units based on tree density, species composition, diameter distribution, and the natural regeneration status of the trees. Silvicultural operations, including harvesting, planting, and tending, are then adapted to the specific conditions of each unit, or forest type. The second methodology involves the marking of trees for harvest. The SSMS predominantly employs single-tree selection harvesting, wherein individual trees are periodically selected and harvested from a broad area.

This chapter presents the fundamental techniques and operational procedures associated with the SSMS as implemented by the UTHF. Initially, the forest types utilized in the system's spatial classification are described. Subsequently, the operational procedures of the SSMS are elaborated. Following this, the operational criteria for tree marking are explained. Finally, the chapter discusses the stand development managed under the selection system.

8.2 Forest Types in the Stand-Based Silvicultural Management System

Forest *stand* is a general term referring to "a contiguous community of trees sufficiently uniform in composition, structure, age, size, class, distribution, spatial arrangement, condition, or location on a site of uniform quality to distinguish it from adjacent communities (Nyland 2007)." Natural forests exhibit diverse and heterogeneous compositions and structures, which would lead to an infinite number of stand types if considered individually. Therefore, the SSMS classifies stands that share the same silvicultural treatments into a single "forest type" for the purpose of silvicultural management planning. This simplifies the process of creating a silvicultural management plan and enables its implementation on a larger, business-oriented scale.

The SSMS originally classified natural forests into three forest types: Selection Cutting Forest, Supplemental Planting Forest, and Clear-Cutting Forest (Takahashi 2001). Selection Cutting Forest encompassed a group of stands where natural regeneration was relatively robust, and selection cutting was considered a viable silvicultural practice. Stands falling under the category of Supplemental Planting Forest were characterized by the potential for successful natural regeneration, but with a low density of juvenile trees. Clear-Cutting Forest included stands where natural regeneration was challenging, and future quantitative and qualitative growth could not be reasonably expected. These classifications were decided onsite by technical staff of the UTHF based on various criteria, including site conditions, stem quality, and the challenges associated with natural regeneration (Takahashi 2001).

Over time, the three initial classifications for the treatment of natural forests have proven to be less suitable as silvicultural operations have continued. Consequently, the categorization of forest types has been periodically revised. In the current 14th education and research plan (2021–2030), forests under the silvicultural management are classified into the following six forest types (The University of Tokyo Forests 2022): (1) Selection Cutting Forest, (2) Young Growth Forest, (3) Sparse Forest, (4) Secondary Forest (established after natural disturbances), (5) Artificially Regenerated Forest, and (6) Restricted Forest. These forest types are further subdivided based on the dominant tree species, their disturbance history, and/or regeneration status. The following section provides an overview of the characteristics and management approaches for each of the forest types, accompanied by relevant photographs.

8.2.1 Selection Cutting Forest

8.2.1.1 Conifer-Dominant Selection Cutting Forest

In this forest type, both natural regeneration of conifers and the presence of marketable standing trees are abundant, allowing for the establishment of a mature conifer-dominated forest stage (Fig. 8.1). The main coniferous species in this forest type is Sakhalin fir (*Abies sachalinensis*), which is a crucial tree species for sustainable

Fig. 8.1 Conifer-dominant selection cutting forest. (Photo by Toshiaki Owari)

selection cutting due to its relatively straightforward natural regeneration process. Sustainable selection cutting practices are feasible, and the forest should be maintained and managed as a mixed forest with conifers as the dominant tree species.

8.2.1.2 Conifer-Dominant Selection Cutting Forest with Poor Regeneration

In this forest type, the natural regeneration of conifers is limited, but there is an abundance of marketable standing trees, and the forest is in a mature stage dominated by conifers (Fig. 8.2). The main coniferous species in this forest type is Yezo spruce (*Picea jezoensis*). Unlike Sakhalin fir, Yezo spruce primarily regenerates on the surface of fallen trees, which presents challenges in ensuring a sufficient number of successor trees. Selection cutting is feasible, but it must be performed carefully to preserve successional trees, which are currently lacking. In forests with low accumulation levels, it might be possible to skip one cutting cycle (either 15 or 20 years in the 14th plan). Natural regeneration is inadequate, and there might be a need for future assisted regeneration operations involving supplemental planting. The forest should be maintained and managed as a mixed forest, with conifers remaining the dominant species.

Fig. 8.2 Conifer-dominant selection cutting forest with poor natural regeneration. (Photo by Toshiaki Owari)

8.2.1.3 Broadleaf-Dominant Selection Cutting Forest

This forest type is a mature stage forest dominated by broadleaf trees, typically with limited natural conifer regeneration, but with an adequate presence of marketable standing trees (Fig. 8.3). The predominant tree species in this forest type include Manchurian ash (*Fraxinus mandshurica* var. *japonica*) in moist sites, Katsura (*Cercidiphyllum japonicum*) in north-facing stream landscapes, and forests dominated by lime and maple (*Tilia* and *Acer* spp.), commonly found on north-facing slopes and flat terrains. In this forest type, tree densities are typically low, and natural regeneration tends to be subpar, with dwarf bamboos (*Sasa* spp.) on drier sites and large ferns on wetter sites. Selection cutting is feasible, but it's important to place greater emphasis on the value growth of the existing high-quality broadleaf trees. Selection cutting is conducted once the trees have reached a usable size. In forests with insufficient natural regeneration of broadleaf trees, it might be possible to skip one cutting cycle. In certain cases, future assisted regeneration operations may be necessary. The forest should be maintained and managed as a mixed forest with broadleaf trees as the dominant species.

Fig. 8.3 Broadleaf-dominant selection cutting forest. (Photo by Toshiaki Owari)

a b

Fig. 8.4 (**a**) Conifer-dominant young growth forest (left) and (**b**) Broadleaf-dominant young growth forest (right). (Photo by Toshiaki Owari)

8.2.2 Young Growth Forest

This forest type is further subdivided based on the dominant tree species, namely, Conifer-dominant Young Growth Forest (Fig. 8.4a) and Broadleaf-dominant Young Growth Forest (Fig. 8.4b). In these forests, marketable standing trees are scarce and the tree population is predominantly in a young growth stage. These forests are characterized by a significant number of small-diameter standing trees (≥1.3 m in height and < 25 cm in DBH), primarily composed of small-diameter conifers (≥500 stems) or broadleaf trees (≥1000 stems) (The University of Tokyo Forests 2022).

This forest type shows good natural regeneration, with a substantial presence of successional trees. However, the volume of medium- and large-diameter trees has decreased, partly due to the damage caused by a typhoon in 1981 and/or repeated selection cutting practices. Selection cutting activities should be temporarily halted in this forest type to allow the existing stands to mature. As successional trees grow, the forest is expected to transition into a state suitable for selection cutting, where either conifers or broadleaf trees will become dominant.

8.2.3 Sparse Forest

This forest type is characterized by a sparse tree cover, with few marketable trees and limited natural regeneration (Fig. 8.5). The forest area faces challenges in terms of forest resource recovery due to low stand density, the proliferation of *Sasa* bamboos on the forest floor, limited natural regeneration, and a scarcity of successional trees. For classification purposes, the number of small-sized conifers and broadleaf trees ($H \geq 1.3$ m and DBH <25 cm) should be less than 300 stems and 500 stems per hectare, respectively (The University of Tokyo Forests 2022). This forest used to be categorized as a "Supplemental Planting Forest," because it required

Fig. 8.5 Sparse forest. (Photo by Toshiaki Owari)

additional efforts to facilitate regeneration. Assisted regeneration measures should be initiated as soon as possible. The current methods involve a combination of natural regeneration supported by site preparation (including surface treatment using heavy machinery and the removal of *Sasa* bamboos) along with low-density conifer planting (refer to Chap. 10). However, it is important to note that canopy homogenization resulting from assisted regeneration operations may increase the forest's susceptibility to wind damage. Therefore, when preparing the site, the minimum number of standing trees should be cut down. In the future, the forest is anticipated to transition into a Selection Cutting Forest.

8.2.4 Secondary Forest

This forest type is a forest that has established following significant natural disturbances, such as past fire events and severe wind-damage. These forests are predominantly dominated by pioneer broadleaf trees. This forest type is further subdivided based on the type of natural disturbance, namely, Secondary Forest after Fire and Wind-damaged Forest. Secondary Forest after Fire (Fig. 8.6a) is a secondary forest that had burned in the 1910s (refer to Chaps. 2 and 13). The primary tree species is Monarch birch (*Betula maximowicziana*), with some areas also dominated by

Fig. 8.6 (**a**) Secondary forest after fire (left) and (**b**) wind-damaged forest (right). (Photo by Toshiaki Owari)

Japanese oak (*Quercus crispula*) and Japanese white birch (*Betula platyphylla*). The lumber market places a high value on pure Monarch birch forests, leading to intensive forest management practices aimed at producing high-quality timber. The forest floor is densely covered with *Sasa* bamboos, and natural regeneration is generally inadequate. In the future, as the number of standing trees decreases and the spacing between the upper layers of trees widens, it will become necessary to undertake assisted regeneration work to maintain the forest's health and productivity.

Wind-damaged Forest, on the other hand, is a secondary forest that is currently undergoing the process of regeneration after experiencing extensive wind-damage, typically caused by recent typhoons (Fig. 8.6b). In this forest type, the damaged trees may be left in place or removed and treated. The goal is to facilitate the conversion of these areas into mixed conifer-broadleaf forests through assisted regeneration, which includes soil surface treatment. If natural regeneration is insufficient, the forest may become sparsely populated and overgrown with *Sasa* bamboo, potentially leading to the need for artificial afforestation through planting of conifer seedlings.

8.2.5 Artificially Regenerated Forest

This forest type is further subdivided based on the measure of artificial regeneration, namely, Plantation Forest and Scarified Site in Natural Forest. Plantation Forest refers to a forest that has been created through artificial planting with the objectives of rejuvenating old Clearcutting Forest and Supplemental Planting Forest, and rehabilitating areas that have suffered damage from forest fires and typhoons (Fig. 8.7a). These forests primarily consist of coniferous tree species. The UTHF primarily features artificial tree species such as Sakhalin fir, Yezo spruce, and Sakhalin spruce (*Picea glehnii*). To guide the forest toward becoming a mixed coniferous forest in the future, the density of standing trees is actively adjusted through thinning practices. Scarified Site in Natural Forest, on the other hand, pertains to forest areas that

Fig. 8.7 (**a**) Plantation forest (left) and (**b**) Scarified site in natural forest (right). (Photo (**a**) by Toshiaki Owari and (**b**) by the UTHF, licensed under CC BY-NC-ND 4.0)

Fig. 8.8 (**a**) Ecosystem conservation forest (left) and (**b**) Difficult-to-operate forest (right). (Photo by Toshiaki Owari)

have been cleared to encourage natural regeneration, typically in areas affected by wind damage in natural forests (Fig. 8.7b).

8.2.6 Restricted Forest

This forest type is further subdivided based on the reasons of management restrictions, namely, Ecosystem Conservation Forest and Difficult-to-operate Forest. Ecosystem Conservation Forest comprises stands that are identified as having a high ecological vulnerability, such as riparian forests, wetlands, alpine zones, and wind-affected areas (Fig. 8.8a). Typically, no active operations are conducted in these forests primarily with a focus on forest conservation. In recent years, there has been an active effort to designate restricted forests, considering river conservation and the safeguarding of water sources. In addition, Difficult-to-Operate Forest refers to a forest type that presents challenges in ground-based forest operation, primarily due to site-specific factors such as steep slopes and extensive rock exposures (Fig. 8.8b).

8.3 Operational Procedures of the Stand-Based Silvicultural Management System

The UTHF has two major Natural Forest Working Groups (NFWGs), categorized based on differences in tree growth conditions and accessibility. Here, the term "working group" refers to a unit of forest organization, comprising an aggregate of compartments managed under the same silvicultural method and cutting cycle (Recknagel 1917). The silvicultural operation method and the cutting cycle period (the time required to complete a cutting cycle when the entire forest area is divided into several management units) are almost identical. The first NFWG is situated in the countryside at low to medium elevations, whereas the second NFWG is positioned in the backcountry at medium to high elevations. In the current 14th education and research plan (2021–2030), the cutting cycle for the first NFWG is 15 years, while for the second NFWG, it is 20 years (The University of Tokyo Forests 2022). The number of management units for each NFWG is set to correspond to the respective years of the cutting cycle.

In the SSMS, natural forests are predominantly managed using the single-tree selection system. This method, known as selection cutting, is a harvesting and regeneration technique that involves harvesting trees at relatively short and consistent intervals, maintaining a relatively small and steady cutting rate. Concurrently, the canopy gaps created by the harvest are filled through regeneration (Ohta et al. 1996). This approach contrasts significantly with clear cutting, where all trees in a specific area are cut down simultaneously, and the site is then regenerated. In Conifer-dominant Selection Cutting Forest, the maximum cutting rate, represented by the volume ratio of growing stock, is set at 16% for the first NFWG and 17% for the second NFWG, respectively (refer to Chap. 2 for historical changes in these rates). Harvesting is conducted while ensuring it falls within the growth rate range.

The work for silvicultural management planning in each area to be treated is executed according to the following procedures: (1) forest type classification survey, (2) plot inventory, (3) in-field review meeting, (4) tree marking, (5) harvesting operation, and (6) harvesting supervision. Each specific area has a dedicated team in charge, currently divided among three main teams: the first survey team, the second survey team, and the civil engineering and harvesting team. As of 2024, each team is staffed with four technical staff members who are responsible for planning and executing the work within the designated area. The following sections provide details and methods for each operation, except for harvesting (refer to Chap. 8).

8.3.1 Forest Type Classification Survey

Under the SSMS, forest surveys initially assess the condition of the managed area and classify forest types using specific criteria and indicators (The University of Tokyo Forests 2022). Before onsite surveying, technical staff consult previous

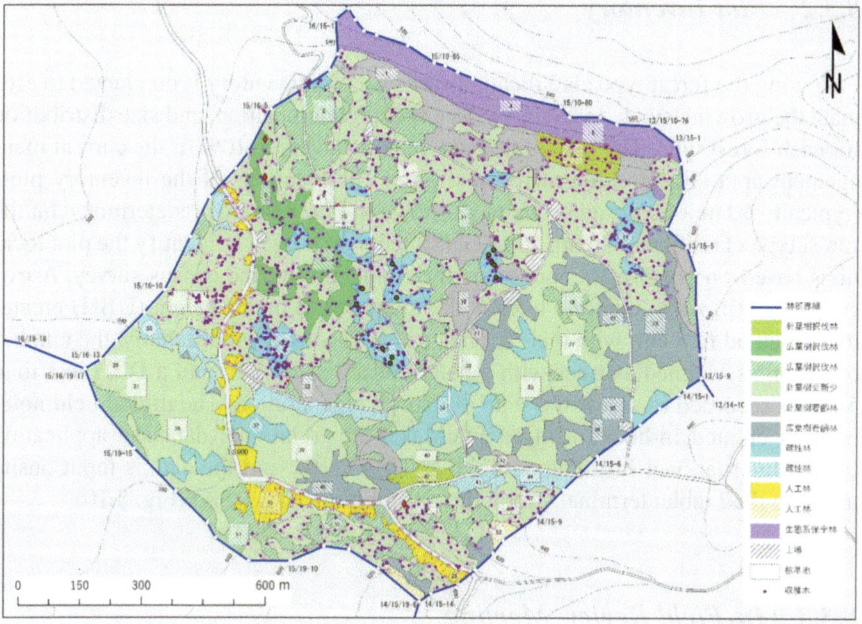

Fig. 8.9 Forest type classification map created using GIS software, displaying the locations of marked trees for harvest obtained via the GNSS receiver. (Source: The University of Tokyo Hokkaido Forest)

maps, management records, aerial photos, and other resources to understand existing forest conditions. Since 2008, the boundaries of forest types have been determined using a laser compass surveying system and, starting in 2012, a global navigation satellite system (GNSS) receiver has been employed for forest type classification surveys (3). This data is then utilized to create forest type classification maps through a geographic information system (GIS) of ArcGIS (Esri, Inc., Redlands, CA, USA) (Fig. 8.9). Additionally, since 2019, the low-cost, dual-frequency real-time kinematic (RTK)-equipped GNSS receiver DG-PRO1RWS (BizStation Corp., Japan) has been used in conjunction with an Android terminal and the field application ArcGIS Field Maps. The positioning accuracy of the system is sufficient for practical use (8), and the simplicity of terminal operations during positioning allows surveying to be a single operator (9).

The UTHF has completed the maintenance of airborne light detection and ranging (LiDAR) data and ortho aerial photographs for the entire forest area. Additionally, recent unmanned aerial vehicle (UAV) aerial photography has been conducted (4). The efficiency of surveying operations has dramatically improved, enabling the delineation of forest types by referencing wide-area spatial information on ArcGIS Field Maps, including forest canopy height models, CS (curvature and slope) topographical maps (Toda 2014), slope volume maps, aerial photographs, and normalized difference vegetation indices (NDVI) (9).

8.3.2 Plot Inventory

Following the forest type classification survey, a plot inventory is conducted to estimate the growing stock, tree density, tree species composition, and size distribution for each forest type. The total plot area typically covers 5–10% of the current management area allocated for each forest type. The locations of the inventory plots (typically 50 m × 50 m square areas, as shown in Fig. 8.9) are predetermined on the forest type classification map, and a GNSS receiver is used to identify the plot locations based on positional coordinates for conducting a tree census survey. A tree caliper is utilized to measure trees with a diameter at breast height (DBH) greater than or equal to 5 cm, with their species names recorded. Additionally, the number of juveniles, defined as trees with a height of at least 1.3 m and a DBH less than 5 cm, is recorded to assess the state of natural regeneration. The digital field notebook is created in-house using a FileMaker Pro relational database application (Claris International Inc., Cupertino, CA, USA) (10), and the data is input onsite using an iPad tablet terminal (Apple Inc., Cupertino, CA, USA) (Fig. 8.10).

8.3.3 In-Field Review Meeting

The team responsible for the current management area consolidates the data from the forest type classification and plot inventory and uses this information to create silvicultural management plans. Following a presentation of the draft plans, in-field

Fig. 8.10 Inventory plot survey conducted using a tree caliper and a tablet terminal. (Photo by Kota Kimura, used with the UTHF's permission, licensed under CC BY-NC-ND 4.0)

Fig. 8.11 In-field review meeting. (Photo by Toshiaki Owari)

review meetings are conducted (Fig. 8.11). During these meetings, staff members discuss how to manage each subcompartment and forest type while physically visiting the management area, ultimately arriving at the final plans. Records are kept to document summaries of the in-field review meetings. Silvicultural prescriptions for the management area are then formulated, and silvicultural practices are subsequently carried out in accordance with these prescriptions.

Participants are provided with ArcGIS Field Maps data by the team in charge and instructed to add the data offline on their tablet terminals. On the day of the in-field meeting, participants check the current location on the field map on their tablet terminals to assess the forest conditions in the field, and confirm that the silvicultural management policy conforms to the actual forest conditions based on the results of the inventory plot survey, and revise the policy as necessary.

8.3.4 Tree Marking

The tree-marking survey is performed within the forest area selected for harvesting according to the silvicultural management plan. This survey involves the careful selection of trees to be harvested. The harvest targets typically include defective

Fig. 8.12 Tree marking. (Photo by Toshiaki Owari)

trees, such as those affected by decay or damaged by fungi or pests, less vital trees, overmature trees, or overstory trees that impede the growth of juvenile trees. Whenever possible, healthy trees are spared from being felled. Tree markers meticulously choose the trees to be cut down by assessing the shape and arrangement of each standing tree (Fig. 8.12).

During the tree-marking survey, a handheld GNSS receiver of the GPSMAP series (Garmin Ltd., Schaffhausen, Switzerland) is carried to record the tree marker's walking trajectory. As in the inventory plot survey, an iPad and FileMaker Pro are used for the survey field book. By synchronizing the field book input with the time of the walking trajectory, the position coordinates of each marked tree can be obtained without operating the GNSS receiver manually. Although the positioning accuracy is inferior to that of the GNSS receiver used for forest classification surveys, it is sufficient for determining the locations of marked trees. The locations of individual marked trees can be displayed on a map, leading to a more efficient search for harvesting trees by standing tree buyers and logging supervision for logging companies.

8.3.5 Harvesting Supervision

In areas where standing trees have been sold, technical staff conduct harvesting supervision works during or after the harvesting operations carried out by logging contractors hired by the standing tree buyers. On-site inspections ensure that the

Fig. 8.13 On-site inspection after logging operations for harvesting supervision. (Photo by the UTHF, licensed under CC BY-NC-ND 4.0)

selected and marked trees have been harvested correctly, that machine trails are properly laid out, and that there are no obstructions caused by the felling or transporting operations. In some cases, compensation may be charged for damaged trees. The presence of local logging contractors with extensive experience working in the UTHF, along with their familiarity with the SSMS, contributes to the careful execution of harvesting operations that minimize forest damage (Fig. 8.13).

8.4 Tree-Marking Criteria

During tree-marking operations, the selection of trees to be harvested should be carried out in a way that enhances forest health and promotes the regeneration and growth of successional juvenile trees (The University of Tokyo Forests 2022). This process aims to increase the growing stock and economic value of the forest without setting specific targets for diameter or accumulation. The primary criteria for tree marking are as follows (Fig. 8.14):

(i) Trees affected by diseases or insects, declining trees with small crowns and sparse foliage, and old, overmature trees that are unlikely to survive until the next cutting cycle.

Fig. 8.14 Tree marking for the single-tree selection system. Trees inhibiting the growth of successional trees (**a** and **c** in the figure) and trees with declining vigor (**b**) should be marked for harvest. (Source: The University of Tokyo Hokkaido Forest)

(ii) Poorly shaped trees inhibit growth of saplings and small- to medium-diameter trees.
(iii) Trees with significant stem defects, such as decay or crookedness.
(iv) Trees that are not expected to grow further or are likely to decline in quality as they increase in diameter.
(v) Thinning to control tree density in areas where the trees are particularly dense.

According to the new initiatives in the 14th plan (The University of Tokyo Forests 2022), the proportion of broadleaf trees marked for harvest will increase in the second NFWG. Specifically, the making of high-quality Japanese oak trees, which meets criterion iv, is attempted in sites managed by the UTHF for direct timber production. In the second NFWG, Yezo spruce trees that meet the tree-marking criteria are intentionally left standing to conserve resources and promote the natural regeneration of the species. When group selection cutting is introduced in parts of Conifer-dominant Selection Cutting Forests with Poor Regeneration or Broadleaf-dominant Selection Cutting Forests, a few surrounding trees that inhibit natural regeneration are marked for harvest, in addition to removing medium- and large-diameter trees of poor quality that meet the tree-marking criteria.

8.5 Stand Development Managed Under the Selection System

At first glance, Fig. 8.15 appears to show an uncultivated forest; however, it has actually undergone six events of selection cutting events over the past 60 years. Trees are marked and harvested approximately every 10–15 years, which affects

Fig. 8.15 A natural forest that has undergone repeated selection cutting in compartment 51. (Photo by the UTHF, licensed under CC BY-NC-ND 4.0)

7–17% of the growing stock (the stem volume of trees that comprise the stand). In 1969, the growing stock of this stand was 301 m^3/ha. By 2018, it had increased to 417 m^3/ha, demonstrating that the growing stock has increased despite six rounds of harvesting, which removed a total stem volume of 201 m^3/ha (see Fig. 8.16). This growth is presumed to result from moderate harvesting practices that enhance the light environment within the stand, thereby promoting tree regeneration and growth. In areas where natural regeneration of conifers, like Sakhalin fir, is successful—particularly on south-facing slopes at lower elevations—sustainable timber production in natural forests is achievable.

Acknowledgments The author would like to thank current and former staff members at the UTHF for their continuous and great efforts managing the forest under the SSMS. The description in this chapter is mainly based on Owari (2014) and Owari (2022), with textual improvements and photo replacements.

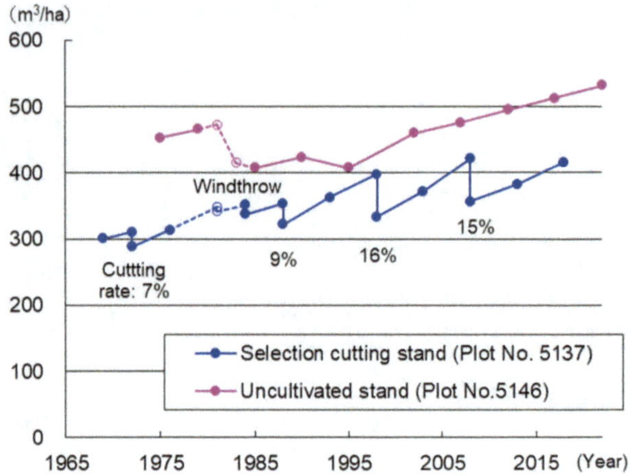

Fig. 8.16 Changes in growing stock in a stand with selection cutting, compared to an uncultivated stand that has not been cut since 1927. The growing stock, shown as an open circle, is estimated from the growth rate of the previous period and the volume of dead trees due to the windthrow in 1981. In the selection cutting stand, salvage logging was conducted after the windthrow. (Source: The University of Tokyo Hokkaido Forest)

References

Nyland RD (2007) Silviculture: concepts and applications, 2nd edn. Prospect Heights: Waveland Press, Long Grove

Ohta T, Kitamira M, Kumazaki M, Suzuki K, Sudo S, Tadaki Y, Fujimori T (eds) (1996) Shinrin no hyakka jiten [Encyclopedia of Forests]. 826pp, Maruzen, Tokyo. (in Japanese)

Owari T (2014) Rinbun Segyo ho wo manabu [learning the stand-based Silvicultural management system]. In: The University of Tokyo forests press (ed.), Jukai wo yuku: Furano, Tokyo Daigaku Enshurin no mori zukuri [traveling through the sea of trees: forest management in the University of Tokyo Hokkaido Forest, Furano]. The University of Tokyo Forests Press, Tokyo, pp 90–112. (in Japanese)

Owari T (2022) DX of natural forest management under the selection system. Sanrin 1656:36–42. (in Japanese)

Recknagel AB (1917) The theory and practice of working plans (forest organization). Wiley, New York

Takahashi N (2001) Rinbun segyo ho: sono kangae to jissen (kaitei ban) [the stand-based Silvicultural management system: its theory and practices (revised Ed.)]. 125pp, Log Bee, Sapporo. (in Japanese)

The University of Tokyo Forests (2022) Education and research plan (2021–2030) of the University of Tokyo Forests, Part 3 Regional forest plans, The University of Tokyo Hokkaido Forest (The 14th period). Miscellaneous Information of The University of Tokyo Forests 64:103–190. (in Japanese). https://doi.org/10.15083/0002002908

Toda K (2014) Topographical interpretation using the CS (curvature and slope) topographical map. Japanese J Forest Environ 56(2):75–79. (in Japanese). https://doi.org/10.18922/jjfe.56.2_75

Chapter 9
Harvesting, Timber Marketing, and Forest Roads

Toshiaki Owari

Abstract This chapter introduces the essential techniques and infrastructures for forest management practices at the University of Tokyo Hokkaido Forest (UTHF), focusing on harvesting, timber marketing, and forest roads. Timber harvesting at the UTHF relied on horses for log transportation until the late 1960s. From the 1960s onward, the UTHF transitioned to using winch and grapple skidders. In 2022, the UTHF further enhanced its timber production system by introducing a forwarder fitted with a grapple crane. Harvested trees at the UTHF are offered either as standing trees or as logs. High-quality hardwood logs command the highest market value and are auctioned at the regional fancy wood auction market. Monarch birch and castor aralia, which are particularly valuable, are primarily used as face veneers for interior woodwork. Forest road construction was actively pursued after the 1960s, resulting in a total road length of 947.1 km and a road density of 41.7 m/ha. Periodic maintenance of forest roads is crucial for the effective management of the UTHF. In August 2016, heavy rains associated with typhoons caused severe damage to these forest roads, and it took 3 years to complete the disaster recovery work.

Keywords Fancy wood auction market · Forest road maintenance · Forest road network · Forwarder · High-quality hardwood timber

9.1 Introduction

Technological advancements, market strategies, and infrastructural management are essential to forest operations. This chapter examines key forest management practices at the University of Tokyo Hokkaido Forest (UTHF) in three areas. First, it outlines the evolution of harvesting methods, from horse-drawn log transportation

T. Owari (✉)
The University of Tokyo Hokkaido Forest, Graduate School of Agricultural and Life Sciences, The University of Tokyo, Furano, Japan
e-mail: owari@g.ecc.u-tokyo.ac.jp

T. Owari et al. (eds.), *Integrative Forest Management and Silviculture*, Ecological Research Monographs, https://doi.org/10.1007/978-981-95-1017-7_9

to a modern forwarder with a grapple crane. Second, it reviews the primary sales methods for harvested trees: standing timber and processed logs, highlighting high-value hardwoods sold at auction. Third, it discusses the development and maintenance of UTHF's 947.1 km forest road network, focusing on construction history, maintenance, and the impact of typhoon damage in 2016.

9.2 Harvesting

In the University of Tokyo Hokkaido Forest (UTHF), while standing tree sales began in 1906, timber production directly by the UTHF started in 1907 (Tokyo Imperial University Forest, Hokkaido 1934). Initially, due to a shortage of workers and a lack of staff members with supervisory experience, logging contractors were hired to perform the direct timber production work. Starting in 1919, the UTHF began to directly employ workers for timber harvesting, primarily compensating them on an individual piece-rate basis (Tokyo University Forests 1950). Felling and processing were performed manually with saws and axes, while horses were used to transport the logs to the landing site (Fig. 9.1). This work was typically seasonal,

Fig. 9.1 Log transportation by horse in 1918, Compartment 103. Logs were skidded across the snow to the landing site, with each horse transporting approximately 2.8 m^3 of logs. (Photo by the UTHF, licensed under CC BY-NC-ND 4.0)

carried out during the winter months. Under the forest colonization system (refer to Chap. 2), settlers were required to participate in harvesting and other forestry activities. Until around 1945, the number of workers assigned to each district was determined by a mandatory quota (Sakaguchi et al. 2001).

A chainsaw was first introduced for tree felling in 1955, and by 1963, nearly all felling and processing operations were conducted by chainsaws. In 1960, a winch skidder, the CT35 (Iwafuji Industrial Co., Ltd., Japan) was first introduced for transportation. Crawler-type winch skidders such as the CT35 and D40 (Komatsu Ltd., Japan) were used, and a wheel-type skidder of FT2 (Mitsubishi Heavy Industries, Ltd., Japan) was also employed for a time. Despite the introduction of skidders, horse-drawn transportation continued simultaneously until the late 1960s. Cut-to-length logging continued until 1981, followed by the half-tree length method in 1972 and full-tree length harvesting after 1990 (Sakaguchi et al. 2001). In 1996, a D4HTS grapple skidder (Shin Caterpillar Mitsubishi Ltd., Japan) (Fig. 9.2) was introduced into the harvesting operations, which were conducted directly by the technical staff of the UTHF (Nitami et al. 1997).

At present, trees selected during the tree marking survey are available for sale either as standing trees or as logs. When sold as standing trees, the responsibility for harvesting operations falls upon the purchaser, typically a private company. Alternatively, for log sales, the UTHF manages harvesting directly or may assign the task to an external contractor. Direct harvesting operations are conducted primarily within managed areas identified as the second Natural Forest Working Group

Fig. 9.2 Transportation of a trunk by a D4HTS grapple skidder (Shin Caterpillar Mitsubishi Ltd.) in Compartment 90, 2007. (Photo by the UTHF, licensed under CC BY-NC-ND 4.0)

(refer to Chapter 8). These operations typically take place from September to November, lasting approximately 3 months (Owari 2022).

Since 2022, UTHF has transitioned its direct timber production system from the tree length method, which utilized a grapple skidder, to the cut-to-length method using a forwarder, despite felling and processing not being fully mechanized. After tree marking for harvest, the marked trees are felled, branches cleared, and the trunks bunched along machine trails by a grapple excavator. These operations are subcontracted to an external contractor, while subsequent processes are directly conducted by the technical staff of the UTHF. Trunks are placed on a log-laying platform in the forest with a PC138US-10 excavator (Komatsu Ltd., Japan) equipped with a GS-90LJV grapple (Iwafuji Industrial Co., Ltd., Japan) and sawn into logs using chainsaws. The logs are subsequently graded according to the Japanese Agricultural Standards (JAS), transported to a landing site using an F801 Forwarder (Kato Works Co., Ltd., Japan) fitted with a grapple crane (Kesla Oyj, Finland) (Fig. 9.3), and stacked by the grapple excavator.

In transportation operations, the machine operator utilizes a GPSMAP 64csx global navigation satellite system (GNSS) receiver (Garmin Ltd., Schaffhausen, Switzerland) to verify the locations of harvested trees, streamline work procedures, and ensure no trees are overlooked during transportation. Simultaneously, the GNSS receiver logs the machine's travel route, enabling the operator to review completed harvesting areas. Aerial photography with unmanned aerial vehicles (UAVs) is also being piloted to preliminarily identify the location of logged trees in the forest prior

Fig. 9.3 Log transportation operation using an F801 Forwarder (Kato Works Co., Ltd.) equipped with a grapple crane (Kesla Oyj, Finland). (Photo by the UTHF, licensed under CC BY-NC-ND 4.0)

Fig. 9.4 Recording log attributes using a tablet devise. (Photo by Kota Kimura, used with the UTHF's permission, licensed under CC BY-NC-ND 4.0)

to transport operations. The technical staff also employs an iPad tablet devise (Apple Inc., Cupertino, CA, USA) with a FileMaker Pro relational database application (Claris International Inc., Cupertino, CA, USA) at the landing site to record daily log yields sequentially, facilitating efficient work progress management (Fig. 9.4) (Owari 2022).

9.3 Timber Marketing

At the UTHF, harvested trees are made available for sale in two forms: as standing trees and as logs. Both forms undergo a bidding process. In the case of standing tree sales, as previously mentioned, the purchasing company is responsible for carrying out the harvesting. For log sales, the logs are prepared by the UTHF, either by its technical staff or through external contractors.

During the 13th planning period (2011–2020), the annual average sales of standing trees amounted 22,037 m^3, with an average unit price of 2011 JPY/m^3. Approximately two-thirds of the standing trees sold originated from natural forests (primarily through selection cutting), while the remaining one-third sourced from plantation forests (largely through thinning). The average unit price of standing trees from plantation forests (2061 JPY/m^3) slightly exceeded that of those from natural forests (1896 JPY/m^3) (The University of Tokyo Forests 2022). The annual average sales volume of logs during the same period stood at 1323 m^3, with an average unit price of 31,551 JPY/m^3. Among these logs, 43% of logs were produced directly by the UTHF staff and sold at an average unit price of 14,692 JPY/m^3.

Since 1996, with the initiation of the 11th management and experiment plan, the UTHF established a dedicated working section for high-value hardwood trees (Tokyo University Forest in Hokkaido 1997). This section encompasses carefully selected hardwood trees managed individually across the UTHF, with a focus on resource management and the production of premium hardwood timber. Emphasis is placed on selling high-quality hardwoods at the fancy wood auction market (The University of Tokyo Forests 2022). Dead or declining trees, including species such as Monarch birch (*Betula maximowicziana*) and castor aralia (*Kalopanax septemlobus*), expected to fetch high prices at auction, undergo assessment based on their projected usability within 15–20 years, before the next silvicultural management cycle (Fig. 9.5). These trees are meticulously cataloged and monitored within a registry of candidate trees earmarked for timber sales (Owari et al. 2016). Upon reaching optimal harvesting maturity, these trees are individually auctioned at the fancy wood auction market in Asahikawa, located approximately 70 km north of the UTHF (Fig. 9.6).

In the 13th planning period, 50–70 m^3 of high-quality hardwood logs were annually auctioned at the fancy wood auction market, achieving an average unit price of 369,207 JPY/m^3 (The University of Tokyo Forests 2022). The highest unit price for monarch birch at the Asahikawa fancy auction market was 2.9 million JPY/m^3 in

Fig. 9.5 A deceased monarch birch (*Betula maximowicziana*) tree in Compartment 80, still available for sale at the fancy wood auction market. (Photo by the UTHF, licensed under CC BY-NC-ND 4.0)

Fig. 9.6 Hardwood logs from the UTHF showcased at the fancy wood auction market in Asahikawa. (Photo by the UTHF, licensed under CC BY-NC-ND 4.0)

1992 (Kasahara et al. 2020). Monarch birch and castor aralia, which are more expensive, are primarily utilized as face veneers for interior woodwork (Okamura et al. 2008).

Japanese oak (*Quercus crispula*) is managed through a specialized registry for candidate trees designated for whisky barrels. This endeavor seeks to efficiently utilize and sustainably manage Japanese oak resources, emphasizing the production of value-added wood products such as barrel timber. The management strategy involves meticulous monitoring via this registry to ensure that each tree fulfills its potential for crafting leak-proof barrels (The University of Tokyo Forests 2022).

9.4 Forest Roads

Following its establishment, the UTHF initially constructed forest tramways before developing the forest road network. The establishment of forest tramways began in 1920 in the Nishitappu district (Ota 1996). By the end of the 5th period (1926–1935), their total length had extended to 50 km (The University of Tokyo Forests 2022). Over time, as the importance of forest roads for automobile traffic increased, the use of horses for farming near the UTHF became impractical for winter logging, leading to a gradual shift to trucks for transporting logs to railway stations and landing sites (Igarashi et al. 1994). In 1955, during the middle of the 7th planning period

(1948–1957), heavy machinery such as bulldozers was first introduced for building forest roads (Anazawa et al. 1988). Concurrently, forest tramways began to be phased out.

At the beginning of the 8th planning period (1958–1967), the total length of forest roads was 33.6 km, with a road density of 1.5 m/ha (The University of Tokyo Forests 2022). During this period, 137.8 km of forest roads were constructed directly by the UTHF staff. Due to labor shortages, extending forest roads was considered critical in the 9th planning period (1968–1980) to improve work efficiency and rationalize operations (The University of Tokyo Forests 2022). An additional 517.7 km of forest roads were constructed in this period (Fig. 9.7), as the construction was strongly promoted by requiring logging companies to build skidding trails along the planned forest road lines when disposing of standing timber (Anazawa et al. 1988; The University of Tokyo Forests 2022). By the end of the 9th period, the total length of forest roads had reached 689.1 km, with a road density of 30.3 m/ha (The University of Tokyo Forests 2022).

Due to the severe typhoon damage in 1981 and subsequent forest restoration efforts, there was little new construction of forest roads until 1985 (Nitami and Anazawa 1997). Since the beginning of the 10th planning period (1986–1995), the UTHF once again became actively involved in constructing forest roads. A total of 213.7 km of forest roads were established in the 10th period, and 38.1 km in the 11th period (1996–2005), with the density of forest roads reaching 39.2 m/ha and 42.0 m/ha, respectively (The University of Tokyo Forests 2022). No forest roads have been constructed since 2004.

Fig. 9.7 Forest road construction using two bulldozers in 1968. Front: CAT D6c; Rear: Komatsu D50A, Compartment 41. (Photo by the UTHF, licensed under CC BY-NC-ND 4.0)

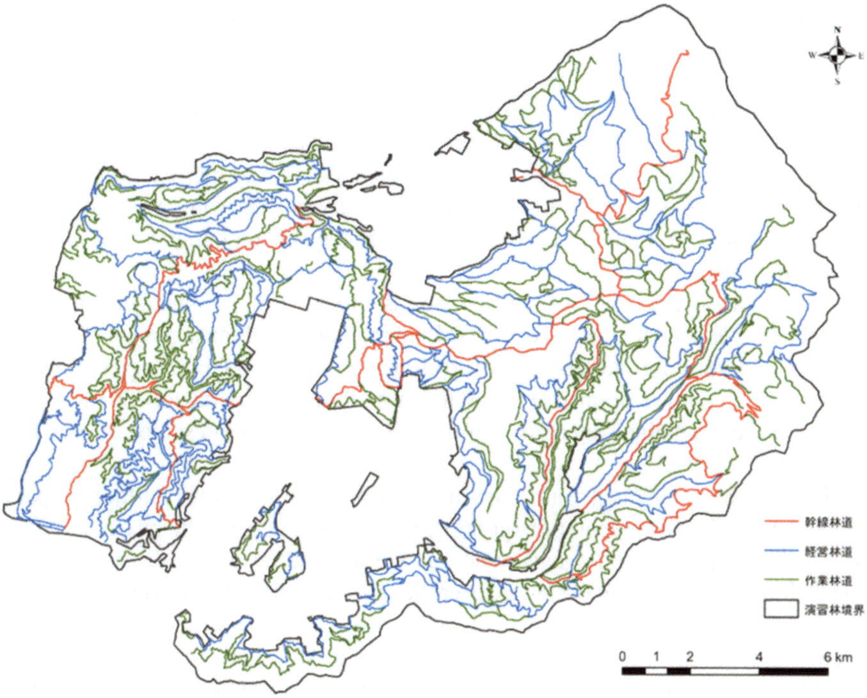

Fig. 9.8 Forest road network of the University of Tokyo Hokkaido Forest. Source: The University of Tokyo Forests (2022)

As of March 2021, the total length of forest roads in the UTHF was 947.1 km, with a road density of 41.7 m/ha (Fig. 9.8). Between 2017 and 2020, 7.4 km of routes were abandoned and subsequently removed from the forest road register (The University of Tokyo Forests 2022). This high-density forest road network, covering the entire forest area of the UTHF, facilitates deep forest access and supports educational, research, and forest management activities.

The forest road network consists of three types of roads (The University of Tokyo Forests 2022). Among these, the main forest roads are crucial as they connect the entire forest area, effectively linking forests to markets and forming the backbone of the network. These roads are permanent facilities, accommodating not only trucks but also large buses and other vehicles. There are 11 main routes totaling 112.6 km. Management roads, branching off from main or public roads into operational areas, serve as key routes for silvicultural management. There are also permanent facilities accessible by trucks year-round, with a total of 81 routes covering 371.0 km. Lastly, operation roads are designed for timber and forestry operations, branching off from management and main roads into enter the forest. They are truck-accessible roads used in each rotation year of silvicultural management and are cost-effective for winter operations, totaling 463.5 km in length.

In the 14th education and research plan of the UTHF, there are four basic guidelines for forest road improvement (The University of Tokyo Forests 2022). Firstly, forest roads lacking safety for forest protection, vehicle operation, and road work should be improved. Secondly, efforts should focus on enhancing the safety of the existing road network and establishing a more intensive silvicultural system. Thirdly, improving the existing road network should aim to increase timber prices for both standing timber and logs. Lastly, flexibility in improving the existing road network should be maintained according to forest types, topography, and management practices.

Periodical maintenance of forest roads is crucial for the management of the UTHF (Potočnik et al. 2005). As of 2024, maintenance and management of the forest road network are handled by two technicians from the Civil Engineering and Production Section, responsible for civil engineering. Approximately 50 km of forest roads are maintained annually, primarily in the managed forest area of each year (Oikawa et al. 2013).

Forest road maintenance involves four main operations: road surface preparation, slope and ditch preparation, grading, and collapse repair (Oikawa et al. 2013). Road surface preparation entails removing grasses and *Sasa* spp. obstructing vehicle traffic, and smoothing out surfaces eroded by rainwater and snowmelt (Fig. 9.9). Slope and ditch preparation involve cleaning grasses, *Sasa* spp., and shrubs from cut slopes and redigging existing ditches. Additional tasks include removing overhanging bushes and tree branches, clearing accumulated soil and sand at cross-drainage

Fig. 9.9 Road surface maintenance performed by a bulldozer. (Photo by the UTHF, licensed under CC BY-NC-ND 4.0)

Fig. 9.10 Forest road disrupted by successive typhoons in August 2016 (Compartment 92). (Photo by the UTHF, licensed under CC BY-NC-ND 4.0)

pipe mouths and outlets, and burying cross-drainage pipes. Grading involves spreading crushed gravel in erosion-prone areas and leveling it to a uniform thickness and width. Collapse repair focuses on restoring forest roads that have collapsed due to early spring snowmelt or flash floods after heavy rains.

In August 2016, heavy rains associated with four successive typhoons caused severe damage to forest roads in the UTHF (Fig. 9.10). The number of identified damaged areas exceeded 100 (The University of Tokyo Forests 2022). The types of damage were categorized as collapse, disruption, and scouring of the traveled road surface, each varying in size and shape (Isozaki et al. 2018). Initially, temporary restoration work was undertaken by UTHF staff and contractors on major forest roads and relatively small areas needing urgent access (Isozaki et al. 2018). Full-scale disaster recovery work commenced in June 2018, one year and ten months after the disaster. Due to the extensive restoration needs across affected areas in Hokkaido, securing contractors took considerable time. The recovery efforts covered 67 damaged sites and were completed in March 2019 (The University of Tokyo Forests 2022).

Acknowledgments The author would like to thank current and former staff members of the UTHF for their continuous and great efforts on harvesting, timber marketing, and building and maintaining forest roads.

References

Anazawa C, Igarashi H, Iinuma T, Inaba B (1988) Road network development in the University of Tokyo Hokkaido Forest: an example of execution in Sakinzawa area during the 10th management period. Proc Conf Test Res Train Tech Staff Univ Tokyo For: 23–33 (in Japanese)

Igarashi H, Anazawa C, Inaba B, Iinuma T (1994) Development and construction work of forest roads in the University of Tokyo Hokkaido Forest. Proc Conf Test Res Train Tech Staff Univ Tokyo For: 97–107 (in Japanese)

Isozaki Y, Oikawa N, Adachi Y (2018) Damage to forest roads caused by successive typhoons in August 2016. Proc Conf Test Res Train Tech Staff Univ Tokyo For: 5–9 (in Japanese)

Kasahara H, Hiroshima T, Okamura K, Inukai H, Fukushi K, Matsui M (2020) The University of Tokyo Hokkaido Forest auctions fancy logs for sale. North For 71:91–95. (in Japanese)

Nitami T, Anazawa T (1997) The development of forest road network and its availability for the harvesting operations. Trans Meet Hokkaido Branch Jpn For Soc 42:256–258. (in Japanese)

Nitami T, Iwamoto S, Fukushi K, Kasahara H (1997) Skidding, the stand disturbance and operational efficiency at the selective harvesting operation in a natural forest. Trans Annu Meet Jpn For Soc 108:493–496. (in Japanese)

Oikawa N, Isozaki Y, Iinuma T (2013) Forest road maintenance operations and data management at the University of Tokyo Hokkaido Forest. Proc Conf Test Res Train Tech Staff Univ Tokyo For: 13–18 (in Japanese)

Okamura K, Inukai H, Owari T (2008) Selling hardwood logs from the University of Tokyo Hokkaido Forest: monarch birch and castor aralia showcased at the Asahikawa Fancy Wood Market. Tree Breed Hokkaido 51:28–30. (in Japanese)

Ota S (1996) Hokkaido Enshurin shinrin kidou kaisetsu no keii [Establishment of forest tramways in the Tokyo University Forest in Hokkaido]. Misc Inform Univ Tokyo For 34:61–158. (in Japanese)

Owari T (2022) DX of natural forest management under the selection system. Sanrin 1656:36–42. (in Japanese)

Owari T, Okamura K, Fukushi K, Kasahara H, Tatsumi S (2016) Single-tree management for high-value timber species in a cool-temperate mixed forest in northern Japan. Int J Biodiv Sci, Ecosyst Serv & Manag 12(1–2):74–82. https://doi.org/10.1080/21513732.2016.1163734

Potočnik I, Yoshioka T, Miyamoto Y, Igarashi H, Sakai H (2005) Maintenance of forest road network by natural forest management in Tokyo University Forest in Hokkaido. Croat J For Eng 26(2):71–78

Sakaguchi T, Iwamoto S, Ihara S, Kimura N, Koike Y, Michigami A, Shibano S (2001) Transition of direct logging operations in the timber production business of the Tokyo University Forest in Hokkaido. Rep Conf Exp Res Train Tech Staff: 46–55 (in Japanese)

The University of Tokyo Forests (2022) Education and research plan (2021–2030) of the University of Tokyo Forests, Part 3 Regional forest plans, The University of Tokyo Hokkaido Forest (The 14th period). Misc Inform Univ Tokyo For 64:103–190. (in Japanese)

Tokyo Imperial University Forest, Hokkaido (1934) The outline of the University Forest in Hokkaido. 46 pp. (in Japanese)

Tokyo University Forest in Hokkaido (1997) The 11th management and experiment plan of the Tokyo University Forest in Hokkaido (1996–2005). Misc Inform Univ Tokyo For 35:1–124. (in Japanese)

Tokyo University Forests (1950) The 50th history of the Tokyo University Forest in Hokkaido. 61pp. (in Japanese)

Chapter 10
Management Practices for Natural Regeneration and Plantations

Satoshi N. Suzuki, Daisuke Sakaue, Susumu Goto, and Toshiaki Owari

Abstract The Stand-based Silvicultural Management System (SSMS) is designed to facilitate the natural regeneration of trees, but natural regeneration often fails even under the SSMS, mostly due to dense dwarf bamboos. Therefore, management practices for assisting the regrowth of forests are necessary for sustainable forest management in Hokkaido. Although soil scarification after clearcutting is one method for assisted natural regeneration (ANR), it often leads to stands with low diversity, dominated by only a few pioneer tree species such as *Betula* spp. Patch scarification without clearcutting can be an effective method for ANR, because small-scale gaps created by selection cutting can establish heterogeneous light environments on the forest floor. The University of Tokyo Hokkaido Forest (UTHF) has also been establishing plantation forests since its early years, but the quantity and methods have changed over time. Before 1980, it was common to establish monoculture plantations, including nonnative species. However, after a major windthrow in 1981, the UTHF has shifted to implementing low-density, polyculture planting

S. N. Suzuki (✉)
The University of Tokyo Hokkaido Forest, Graduate School of Agricultural and Life Sciences, The University of Tokyo, Furano, Japan

Nakagawa Experimental Forest, Field Science Center for Northern Biosphere, Hokkaido University, Otoineppu, Japan
e-mail: snsuzuki@fsc.hokudai.ac.jp

D. Sakaue
The University of Tokyo Chichibu Forest, Graduate School of Agricultural and Life Sciences, The University of Tokyo, Chichibu, Japan
e-mail: sakaue@uf.a.u-tokyo.ac.jp

S. Goto
The University of Tokyo Tanashi Forest, Graduate School of Agricultural and Life Sciences, The University of Tokyo, Nishitokyo, Japan
e-mail: gotos@uf.a.u-tokyo.ac.jp

T. Owari
The University of Tokyo Hokkaido Forest, Graduate School of Agricultural and Life Sciences, The University of Tokyo, Furano, Japan
e-mail: owari@g.ecc.u-tokyo.ac.jp

T. Owari et al. (eds.), *Integrative Forest Management and Silviculture*,
Ecological Research Monographs, https://doi.org/10.1007/978-981-95-1017-7_10

with native conifers, aiming for the afforestation of conifer-broadleaf mixed forests, which are thought to be more resistant to natural disturbances. Furthermore, efforts to improve materials used for plantations have been made through introduction breeding, selective breeding, and cross-breeding.

Keywords Assisted natural regeneration · Soil scarification · Plantation · Forest tree breeding

10.1 Assisted Natural Regeneration

Although the Stand-based Silvicultural Management System (SSMS) aims to facilitate natural tree regeneration, the process is often insufficient even under the SSMS. Dense dwarf bamboo suppresses tree recruitment, resulting in few advanced regenerations on forest floors where bamboo is prevalent (see Fig. 5.4 in Chap. 5). Without adequate advanced regeneration, selection cutting frequently leads to irreversible canopy openings. Additionally, severe natural disturbances can devastate the forest canopy, further enabling bamboo to dominate the forest floor. These open sites, with minimal regeneration, are expected to take a very long time to recover to a closed canopy forest without intervention. Therefore, silvicultural practices are essential to support forest regrowth and ensure sustainable forest management.

In Hokkaido, one strategy for regrowth in degraded stands is conversion to plantation forests. This involves clearing the stands and removing the understories, followed by planting single-species seedlings at a high density (2000–3000 seedlings ha^{-1}). In well-managed plantations, invaded unplanted trees are removed and the planted trees are thinned to regulate density. These forests are typically harvested by clearcutting, after which afforestation is repeated. Such practices facilitate management and predictability of growth and harvesting, making this approach the most common in Japan. However, even-aged monocultures are structurally and functionally homogeneous, making them vulnerable to disturbances like windstorms and insect outbreaks (Liu et al. 2018; Dobor et al. 2020).

An alternative strategy is assisted natural regeneration (ANR), which combines active planting with passive restoration (Chazdon et al. 2022). ANR accelerates natural successional processes by removing barriers to regeneration, such as soil degradation, competition with weedy species, and recurrent disturbances. This approach involves various procedures, including planting native species, scarifying soil to promote regeneration from seeds, and managing vegetation through weeding and bush clearing.

In Hokkaido, soil scarification following clearcutting is a traditional ANR technique used to remove dwarf bamboo and encourage natural tree regeneration (Ito et al. 2018, Fig. 10.1). However, scarified areas are often dominated by pioneer species like birches. These birch stands tend to be structurally homogeneous and have

Fig. 10.1 Soil scarification after clearcutting. (Photo by Satoshi N. Suzuki)

low species diversity. Without thinning, they frequently consist of overcrowded, slender stems, making them unsuitable for timber production.

Soil scarification without clearcutting has also been attempted as an ANR method in Hokkaido. In the Uryu Experimental Forest of Hokkaido University, where forest floor soil scarification was conducted in September 2000, Japanese white birch (*Betula platyphylla* var. *japonica*) was the most abundant species 2 years after scarification (Yoshida et al. 2005). However, conifers, including Sakhalin fir (*Abies sachalinensis*) and Sakhalin spruce (*Picea glehnii*), became the most dominant 6 years after scarification, and saplings of various tree species were present (Harada et al. 2009). A similar trial in a national forest in the northern Hokkaido revealed that both pioneer and late successional species could regenerate at medium levels of light, approximately 30% of the relative photosynthetic photon flux density (Tanimura and Yamasaki 2020). These results suggest that soil scarification without clearcutting facilitates the regeneration of various species with a range of shade tolerances, as small-scale gaps generated by selection cutting create heterogeneous light environments. Therefore, soil scarification without clearcutting can be an effective method for ANR in Hokkaido.

At the University of Tokyo Hokkaido Forest (UTHF), a variety of ANR techniques have been tested. Although ANR practices were limited before the start of the SSMS in 1958, the initial SSMS presupposed that ANR practices should be actively conducted in stands with little regeneration, specifically the "Supplemental

Planting" stand in the original stand classification of the SSMS (see Chap. 8). During the early era of the SSMS (1958–1960), the UTHF actively planned to implement supplementary planting. In the Supplemental Planting Stand, planting was carried out in canopy gaps covered by dwarf bamboos. Conifer seedlings, mostly Sakhalin fir, were also planted around cut stumps after single-tree selection harvests. However, due to labor shortages, these supplementary planting activities, which were conducted manually without the aid of machinery, were discontinued after a few years.

Since the late 1960s, with the development of machinery for scarification, the area of scarification and planting has increased. Relatively large forest stands were cleared and planted, leading to an increase in monoculture plantations in the 1960s and 1970s. Furthermore, after a severe typhoon disturbance in 1981, large-scale afforestation efforts were undertaken (see Chap. 12). At suitable plantation sites, Sakhalin fir and Sakhalin spruce seedlings were planted in monocultures. At less suitable sites, only scarification or seeding after scarification was carried out, partly due to a shortage of seedlings. After the 5-year reforestation period following the 1981 typhoon, afforestation practices involving machinery scarification and conifer planting continued. Although the seeding of conifers and some broadleaves, such as Japanese oaks (*Quercus crispula*) and birches, was partially attempted, most were unsuccessful. As a result, seeding practices have not been conducted in the UTHF since the 1990s.

Since the 1990s, seedlings have been planted at low density, at 1000–1500 seedlings per hectare, which is much lower than the typical planting density for plantation forests in Hokkaido, approximately 2000–3000 seedlings per hectare. High-density planting effectively inhibits nonplanted species from invading the planted area, resulting in a pure monoculture. In contrast, low-density planting aims to reduce the cost of planting and thinning and to facilitate the natural regeneration of broadleaf trees.

10.1.1 Studies on Assisted Natural Regeneration

In the late 1970s, scarification experiments were carried out at the UTHF (Plot Nos. 5501–5512). In 1979, the forest floor in 12 plots, with areas ranging from 0.96 ha to 2.26 ha (the area for Nos. 5504 and 5505 are unknown), was scarified by a bulldozer (D50A; Komatsu, Tokyo, Japan) with a one-tooth attachment after clearcutting of natural mixed forests (Hatano 1984). Scarification using this type of bulldozer created concave and convex microtopographies, which were expected to facilitate the natural regeneration of diverse species, including conifers. These experiments appeared to be successful; mixed stands dominated by Yeso spruce (*Picea jezoensis*) and Erman's birch (*Betula ermanii*) developed at high-elevation sites, Sakhalin fir and Painted maple (*Acer pictum*) at mid-elevation sites, and Monarch birch (*Betula maximowicziana*) at low-elevation sites (Nakagawa et al. 1994; Goto et al. 2010, Fig. 10.2).

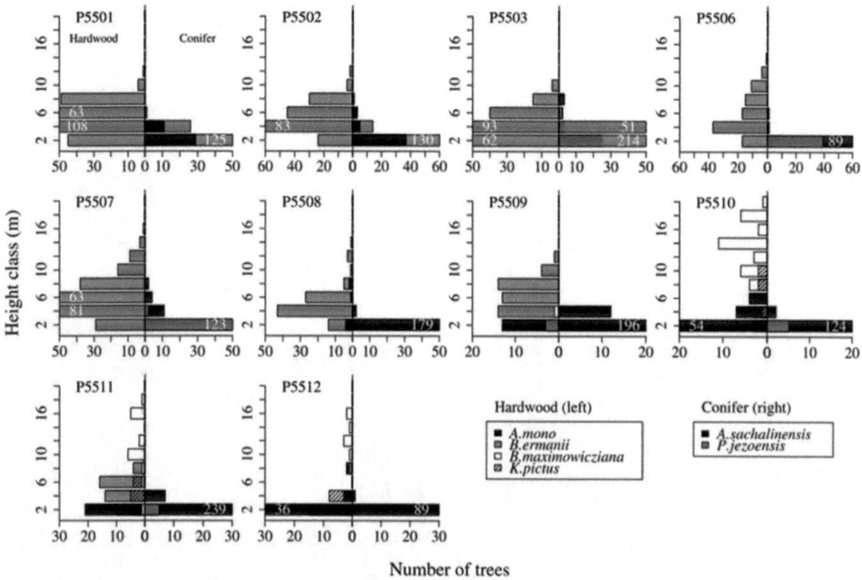

Fig. 10.2 Height distributions of six major species regenerated after 26 years of scarification at the experimental scarification sites in the UTHF. Each panel indicates a site. P5501–5503 and 5506 at 740–760 m in elevation; P5507–P5509 at 615–680 m; P5510–5512 at 420–445 m. *A. mono* refers to *Acer mono* (a synonym of *A. pictum*), and *K. pictus* refers to *Kalopanax pictus* (a synonym of *Kalopanax septemlobus*). (Reproduced from Goto et al. 2010, with the publisher's permission)

However, similar scarification trials conducted after a severe typhoon disturbance in 1981 appeared to be unsuccessful (refer to also Chap. 13). Most of the scarified area was dominated by Erman's birch or Japanese white birch (*B. platyphylla* var. *japonica*) (Fig. 10.3), with few conifers regenerated within the scarified area. In 2001, scarification trials were also conducted to facilitate the natural regeneration of Monarch birch. For several years after scarification, seedlings of many tree species, including Monarch birch, were established (Goto and Tsuda 2007). However, more than 10 years after scarification, the area was covered by dwarf bamboo and shrubs such as *Aralia elata*, with only a few Monarch birch trees observed (Goto et al. 2020). Although the silvicultural treatments used in these trials were similar to those in the experiments that started in 1979, most of them seemed to have failed.

Although conifers and economically valuable broadleaves successfully regenerated in the 1979 scarification experiment, why did they not regenerate after the disturbance caused by the large typhoon in 1981? One possible reason for this difference would be the disparity in seed supply. First, 1979 was the masting year for Yezo spruce and birch species. Additionally, the scarified areas in the 1979 experiment were relatively small and surrounded by the remaining mature trees, which contributed to the seed supply in the scarified area. In contrast, large-scale scarification and salvage logging were carried out in the area disturbed by the 1981 typhoon,

Fig. 10.3 A 42-year-old Erman's birch (*Betula ermanii*) stand regenerated at a site scarified after windthrow in 1981. (Photo by Satoshi N. Suzuki, May 29, 2024)

resulting in few mature trees remaining to supply seeds. Consequently, even-aged stands of *Betula* species could be established after the 1981 typhoon due to their high seed dispersal ability and/or persistence as buried seeds.

Light conditions are another important factor in determining the success of forest tree regeneration. A small-scale clearcutting and scarification experiment of 20 × 20 m areas in a plantation of eastern white pine (*Pinus strobus*) was conducted to promote the natural regeneration of native tree species from surrounding remaining native trees (Plot No. 5522, Takahashi et al. 2023). In these scarified areas, not only birch species but also Sakhalin fir regenerated, especially in the southern part of the scarified areas. However, fewer Sakhalin fir regenerated in the northern part. This could be because, on the north side, where sunlight directly reaches the ground, dryness and light inhibition can affect the establishment of Sakhalin fir, and the growth of light-demanding tree species and herbaceous vegetation limits the growth of Sakhalin fir.

These findings affirm the effectiveness of employing both small-scale harvesting and scarification as a silvicultural method to promote the natural regeneration of various species. The modest canopy openings created through small-scale harvesting ensure the presence of mother trees for seed supply. Additionally, this silvicultural approach mitigates light inhibition and prevents the overgrowth of light-demanding tree species and weeds on the forest floor following scarification.

Fig. 10.4 Small-scale scarification following group selection cutting. (Photo by Satoshi N Suzuki, May 29, 2024)

In this context, beginning in 2020, the UTHF initiated a trial of scarification after group selection cutting, where each canopy opening was limited to less than 0.04 hectares, to facilitate the natural regeneration of a diverse range of native tree species in the selection cutting stand (Fig. 10.4). This method is expected to enable sustainable timber production in forests under selection system, even those with dense dwarf bamboos, by promoting natural regeneration without large-scale canopy openings.

Another ANR method in selection cutting stands is supplemental planting around cut stumps. This method was originally applied in the early era of SSMS as a supplementary planting technique in areas with little regeneration in the Selection stand and Supplementary planting stand (Takahashi 1951; Shibata 1981). After selection cutting, several seedlings (typically three or four seedlings per stump) were planted around the cut stump. This was a cost-effective method before the introduction of machinery scarification because it required fewer seedlings than planting after clearcutting, and weeding was not necessary. However, with the introduction of machinery scarification and increasing labor costs, this method did not last long. Recently, this method has been revisited due to the increased recognition of the disadvantages of large-scale scarification and monoculture planting. Successful survival and growth of planted seedlings around cut stumps has also been reported (Ishibashi et al. 2010).

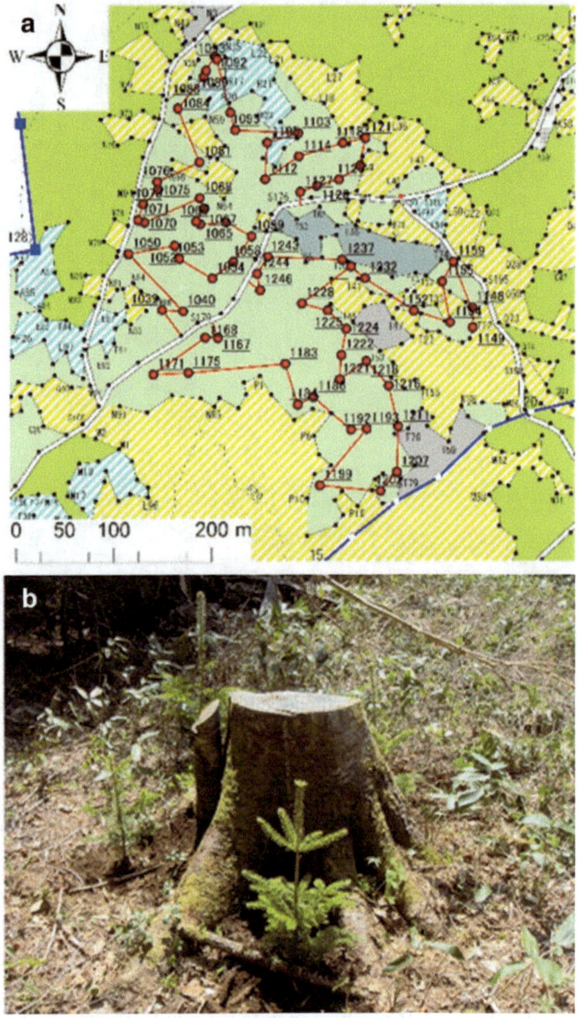

Fig. 10.5 (**a**) The locations of cut stumps (red circles) and a route for the planting procedure (red line). (**b**) Planted seedlings of Sakhalin fir (*Abies sachalinensis*) around a cut stump. (Reproduced from Owari et al. 2015, photo by Toshiaki Owari, May 20, 2014)

Furthermore, recent advancements in geographic information system (GIS) and global navigation satellite system (GNSS) techniques have enabled high-precision mapping of cut stumps and the routing of planting procedures (Owari et al. 2015; Fig. 10.5a). In 2014, UTHF started trials in which seedlings were planted around cut stumps (Fig. 10.5b). The survival rate of the planted Sakhalin fir seedlings was 80–92% after 3 years and 49–70% after 7–8 years after planting (Owari et al. unpublished data). The planted seedlings reached a height equivalent to or greater than the mean height of dwarf bamboo culms (102–132 cm) 7–8 years after planting. Therefore, this method can be regarded as an effective for ensuring the regeneration in selection cutting stands.

Forest grazing experiments were implemented as a silvicultural strategy to control *Sasa* dwarf bamboos and facilitate natural regeneration in 1940–1950s. Overgrazing experiments were initiated in Higashiyama in 1948, covering an area of 0.5 hectares, and in Rokugo in 1949, covering 2.6 hectares (Takahashi 1952). In Higashiyama, 5–8 small livestock, such as goats, sheep, and pigs, were grazed annually during the summer from 1948 to 1952. In Rokugo, 5–8 cows were grazed during the summer from 1949 to 1951. Four years of forest grazing in Higashiyama successfully eradicated *Sasa* dwarf bamboo but nearly eliminated other understory vegetation, except for unpalatable species. Conversely, in Rokugo, *Sasa* dwarf bamboos almost completely vanished from the forest floor, and grazing for more than 3 years resulted in increased numbers of plants of certain tree species and herbaceous plants, such as *Impatiens noli-tangere* (touch-me-not balsam, Fig. 10.6). Although this type of forest grazing activity was not sustained beyond 1970 in the UTHF, the effectiveness of forest grazing has been reported in northern Japan (Yasue 2000; Fukuda et al. 2009).

10.2 Plantation

10.2.1 History of Plantation Forests

In the first decade after the establishment of the UTHF, plantations were implemented at a relatively small scale (Fig. 10.7). The first plantation was established in 1907. During this time, due to the lack of fully developed methods for raising seedlings of native species, introduced species such as Norway spruce (*Picea abies*), Japanese larch (*Larix kaempferi*), and Scots pine (*Pinus sylvestris*) were planted. After large-scale severe fires in the 1910s (see Chap. 12), extensive afforestation practices were implemented, mostly with Norway spruce. Although 70% of the Norway spruce plantations were not successful, the remainder of the Norway spruce plantations have grown into old-growth plantations over the course of a century (Fig. 10.8). From the 1920s onward, plantation efforts were not as intensive. However, the area of plantations began to increase significantly after the late 1950s. Between 1955 and 1975, Sakhalin fir, a native conifer, as well as introduced species such as eastern white pine and larches were planted. Hardwoods were also planted intensively on a trial basis during this period. Although the annual plantation area slightly decreased, due to a lack of labor and increased cost in the late 1970s, the plantation area sharply increased following the typhoon that struck the UTHF in 1981 (see Chap. 13). At this time, two native conifers, Sakhalin fir and Sakhalin spruce, were planted in monoculture. Monoculture plantation of the two conifers continued until around 2000. As methods for growing seedlings of Yezo spruce were developed in the late 1990s (see Chap. 11), the number of plantations of Yezo spruce increased in the early 2000s.

Fig. 10.6 (**a**) *Sasa* dwarf bamboo densely covered the forest floor before the forest grazing experiment in Rokugo. (**b**) Population of western touch-me-not (*Impatiens noli-tangere*) after 3 years of forest grazing in Rokugo. These images were processed and enhanced for higher resolution using Photoshop® (Adobe Inc.). (Modified from Takahashi 1952, with the publisher's permission)

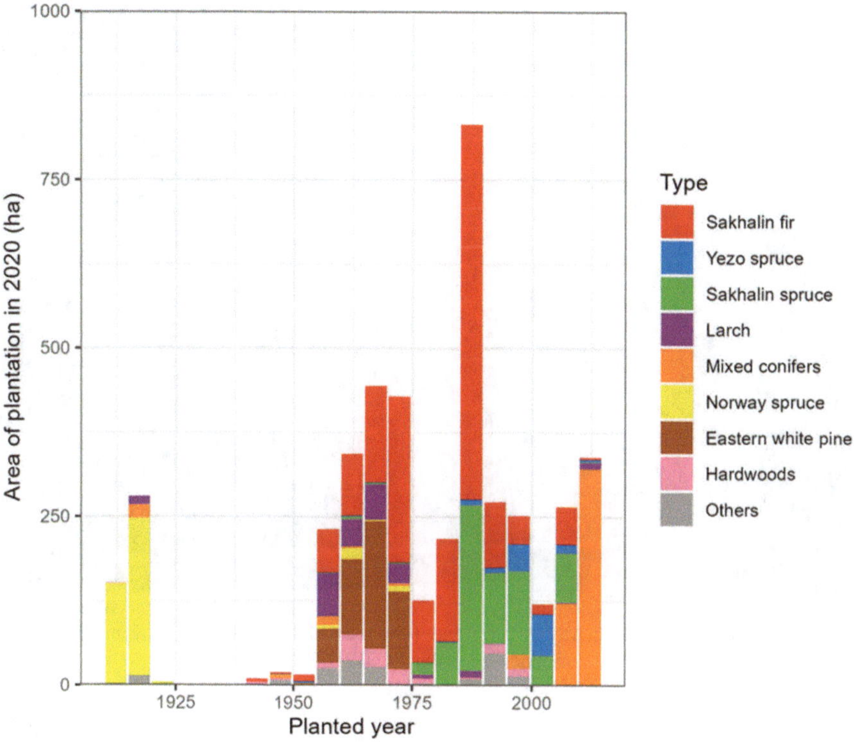

Fig. 10.7 The area and planting year of plantations in the UTHF in 2020. (Data source: UTHF)

After 2005, the plantation strategy shifted to polyculture plantations with a combination of the three native conifers (Fig. 10.9). The planting density was intentionally kept relatively low, as mentioned earlier, to reduce the cost of planting and thinning and to facilitate the invasion of broadleaf plants. This approach aimed to promote the development of mutispecies conifer-broadleaf mixed forests, which are considered more resilient to natural disturbances. In the 2010s, the majority of plantations have adopted this polyculture strategy (The University of Tokyo Forests 2022).

As an exception, larch monocultures continued to be newly afforested on 2 ha annually (The University of Tokyo Forests 2022). For these larch plantations, a hybrid breed of *Larix gmelinii* var. *japonica* × *Larix kaempfeli*, known as "Cleanlarch" has been utilized (Fig. 10.10). Cleanlarch, registered in 1994, exhibits a high growth rate comparable to that of Japanese larch, along with enhanced resistance to mouse damage and greater wood density.

Fig. 10.8 A 112-year-old Norway spruce stand. (Photo by Makoto Ishiguro, November 22, 2023, used with the UTHF's permission, licensed under CC BY-NC-ND 4.0)

Fig. 10.9 A mixed conifer plantation. (Photo by Satoshi N. Suzuki, August 31, 2021)

Fig. 10.10 A 13-year-old Cleanlarch plantation. (Photo by Makoto Ishiguro, October 18, 2023, used with permission from the UTHF, licensed under CC BY-NC-ND 4.0)

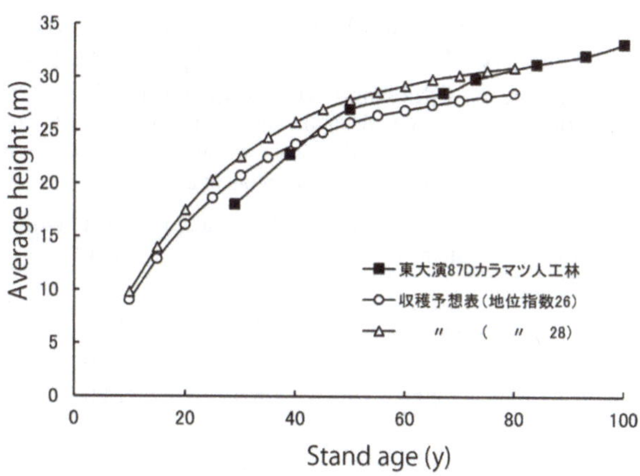

Fig. 10.11 Trajectory of mean tree height in the old larch plantation established in 1908 (black squares). Predicted growth trajectories for stands with site indices of 26 (open circles) and 28 (open triangles), based on the local yield table, are also shown. (Modified after Isozaki et al. 2008, with the publisher's permission)

10.2.2 Experimental Studies of Plantations

10.2.2.1 Reference Forests

Old plantations that were established in the early period of the UTHF in the 1900s and 1910s were registered as "Reference forests." One of the oldest plantation forests in the UTHF is a Japanese larch plantation planted in 1908 (Kurahashi et al. 2001, Isozaki et al. 2008, Karthigesu et al. 2023). This is one of the oldest larch plantations in Hokkaido. Long-term records of these old larch plantations suggest that larch plantations can continue to grow even after reaching 100 years of age (Fig. 10.11).

10.2.2.2 Stand Density Control Experiments

Several stand density control experiments were established in the 1950s and 1960s. Stand density control plots with initial densities of 1500, 3000, 6000, and 15,000 stems ha^{-1} of Sakhalin fir (Plot No. 3042) and eastern white pines (Plot No. 3043) were established in 1969. Although the growing stock in 2003 (35 years old) was highest in plots with 6000 stems ha^{-1} (Table 10.1), a number of naturally regenerated species invaded the lower density plots (especially the 1500 stems ha^{-1}, Fig. 10.12), resulting in the highest species diversity in the lower density plots (Matsui and Okamura 2004). This has important implications for the restoration of a conifer-broadleaf mixed forest. Another stand density control experiment of Sakhalin fir was established in 1969 (Plot No. 3041).

Stand density control plots of Japanese white birch were also established in 1958, with initial stand densities of 2000, 3000, and 4000 stems ha^{-1} (Iwamoto and Nishida 1963; Iwamoto et al. 1970; Hirata et al. 1996). Interestingly, the stand stem density of the plot with an initial density of 4000 stems ha^{-1} steeply decreased until 2012 (at 54 years of age) and became almost equivalent to that of the plot with an initial density of 2000 stems ha^{-1} (Fig. 10.13). Furthermore, the archived stand growing stock of the stand with an initial density of 2000 stems ha^{-1} exceeded that of the stand with an initial density of 4000 stems ha^{-1}. Additionally, the plots with initial densities of 3000 and 4000 stems ha^{-1} were severely damaged by heavy snow packing in 2013 (refer to Chap. 13). As a result, stands with lower initial densities had greater biomass and stem density in 2020 (at 62 years of age), suggesting that

Table 10.1 Survival rate, mean diameter and height of trees, and stand growing stock of Sakhalin fir (*Abies sachalinensis*) and eastern white pine (*Pinus strobus*) stands with different initial densities in the stand density control plots at age 35

Species	Initial density (stems ha^{-1})	Survival rate (%)	Mean DBH (cm)	Mean height (m)	Growing stock (m^3 ha^{-1})
Sakhalin fir	1500	65.4	19.4	17.6	244.3
	3000	44.4	16.9	15.8	237.3
	6000	44.2	15.6	18.4	385.0
	12,000	25.8	14.0	17.5	339.6
Eastern white pine	1500	48.9	27.5	20.1	386.1
	3000	32.8	26.1	20.3	458.1
	6000	22.9	22.8	19.6	473.1
	12,000	14.6	19.6	18.8	420.6

Translated after Matsui and Okamura (2004)

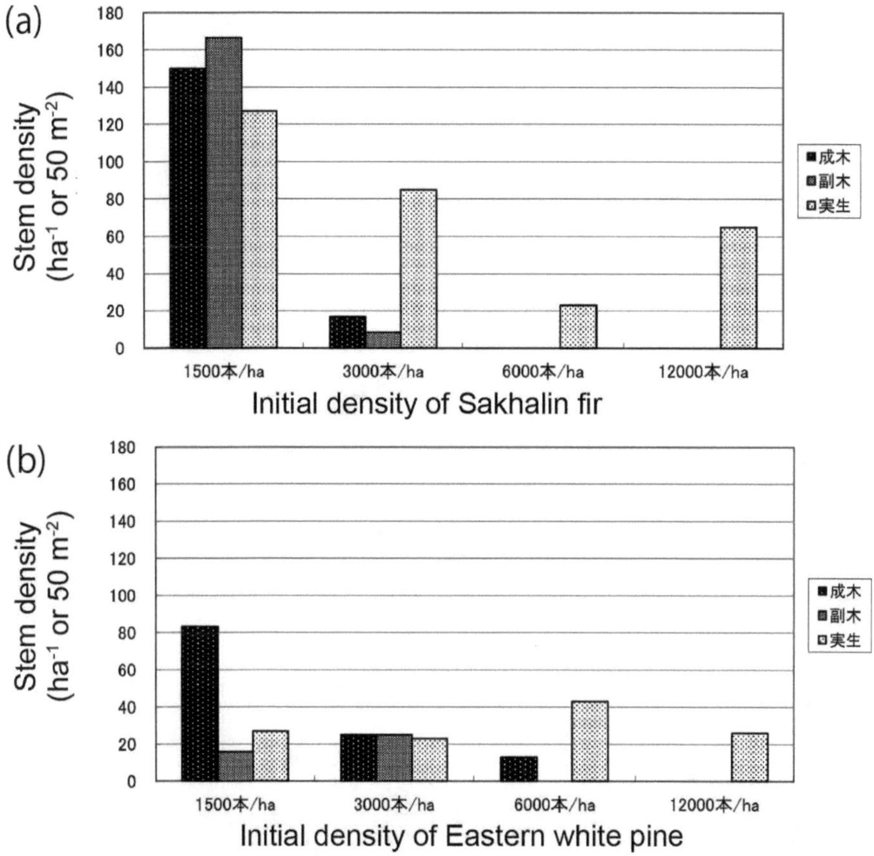

Fig. 10.12 The number of naturally regenerated trees in (**a**) Sakhalin fir and (**b**) eastern white pine stands with different initial densities at age 35. Black bars represent the density of stems larger than or equal to 5 cm in DBH (ha^{-1}); dark gray bars show the density of trees smaller than 5 cm in DBH but taller than or equal to 1.3 m in height (ha^{-1}); light gray bars indicate the density of trees smaller than 1.3 m in height (50 cm^{-2}). (Modified after Matsui and Okamura 2004, with the publisher's permission)

density control, such as thinning, is necessary not only for productivity but also for resistance to disturbances such as snow packing.

10.2.2.3 Density-Control Experiment of Mixed Sakhalin Fir and Japanese Larch Plantations

An intensive experiment was initiated in 1956 to assess the effects of mixing and stand density of Sakhalin fir and Japanese larch (Plot No. 3011, Iwamoto and Nishida 1968; Yamamoto 1992, 1993). Mixed stands were established, combining Sakhalin fir with densities of 1000, 2000, 3000, and 4000 stems per hectare and

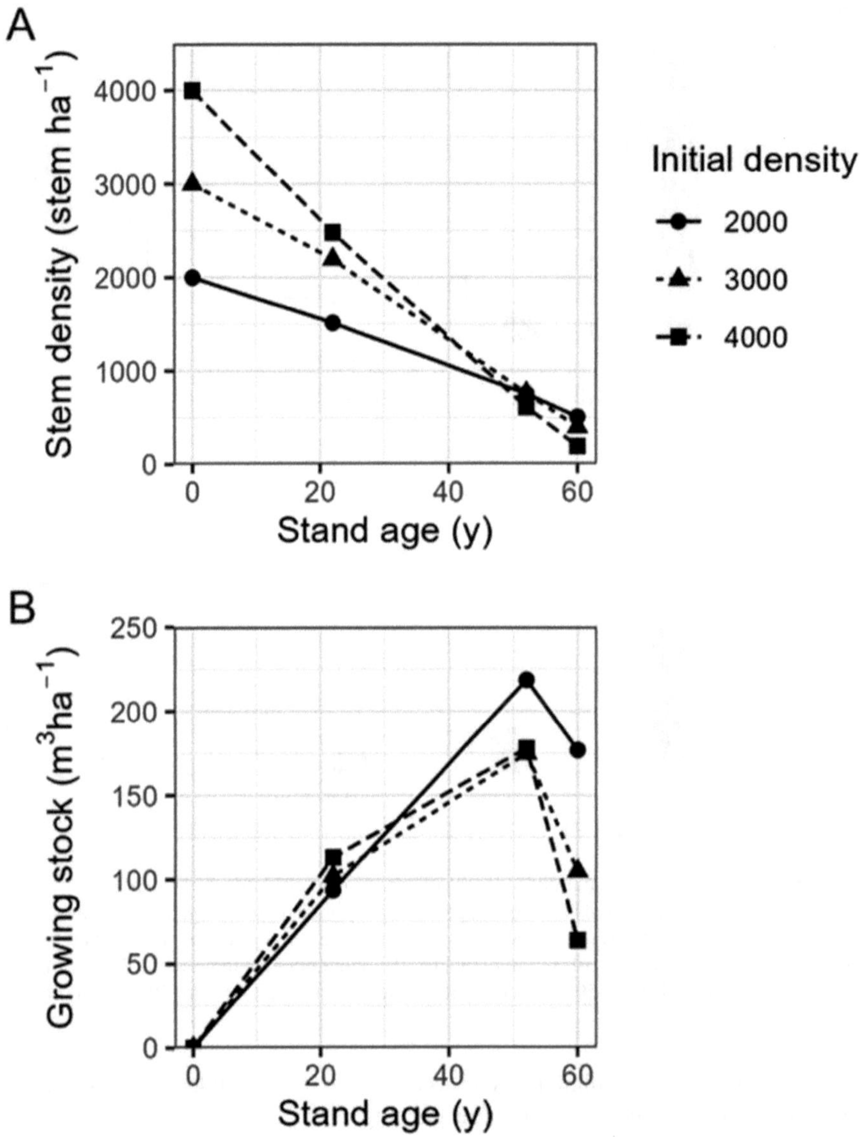

Fig. 10.13 Changes in stem density (**a**) and growing stock (**b**) in the density control experiment of Japanese white birch (*Betula platyphylla* var. *japonica*). (Data source: UTHF)

Japanese larch with densities of 500, 1000, and 1500 stems per hectare. Additionally, monocultures of Sakhalin fir with 3000 stems per hectare and Japanese larch with 1000, 2000, and 3000 stems per hectare were set up. Notably, mixed stands of all combinations exhibited nearly equivalent growing stock at 34 years of age, surpassing that of the monoculture Sakhalin fir plantation (Yamamoto 1992, Fig. 10.14).

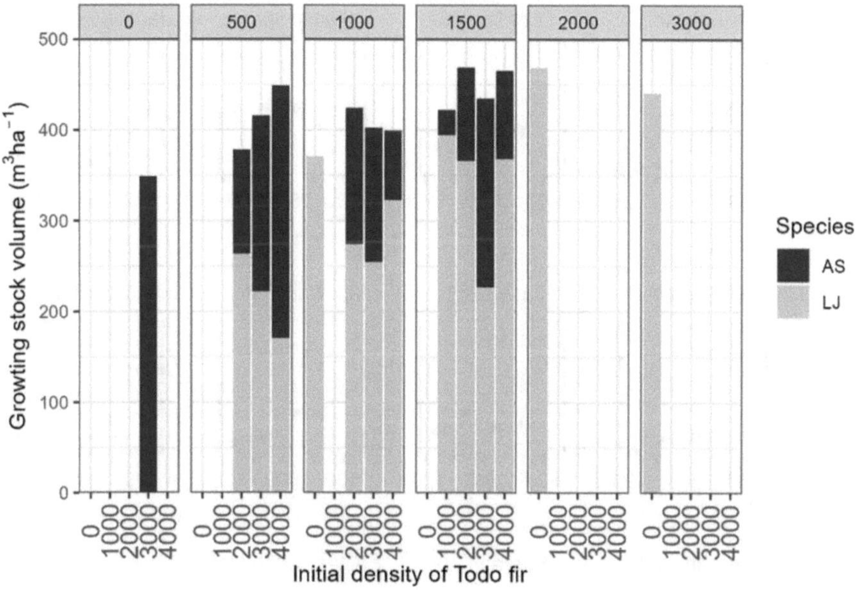

Fig. 10.14 Growing stock in the density control mixed plantation experiment at 34 years of stand age. The value in the header of each panel indicates the initial density of Japanese larch. The black and gray bars indicate the stock volumes of Sakhalin fir (AS) and Japanese larch (LS), respectively. The average of two replicate plots whose values were reported in Yamamoto (1992) is shown

Moreover, in cases where the planting density of Japanese larch was sufficiently low in the mixed stands, the productivity of Sakhalin fir experienced only a modest 20% reduction compared to that of the monoculture. This study implies that incorporating a low-density mixture of Japanese larch into Sakhalin fir plantations enhances overall stand productivity, ensuring the sustained productivity of both fir and larch.

Although several experimental plots of mixed plantations were established until 1960, including Eastern White Pine-Japanese White Birch mixed plantations (Plot No. 3026), Sakhalin Fir-Korean Larch mixed plantations (Plot No. 3027), and Eastern White Pine-Korean Larch mixed plantations (Plot No. 3028), few analytical studies have been performed on these mixed plantation experiments.

10.2.2.4 Environmental Adaptability Experiment

Experimental plantations along the elevation gradient in the UTHF were established to assess the climatic suitability of plantation species. In 1959, nine conifers and two broadleaf species were planted at four elevations: 530, 730, 930, and 1100 m above sea level (a.s.l.). The species included Sakhalin fir, Yezo spruce, Sakhalin spruce, Japanese larch, Kurile larch, F1 hybrid of Kurile larch × Japanese larch, Korean larch (*L. gmelinii* var. *olgensis*), F1 hybrid of Korean larch × Japanese larch, Siberian larch (*Larix sibirica*), American larch (*Larix laricina*), eastern white pine,

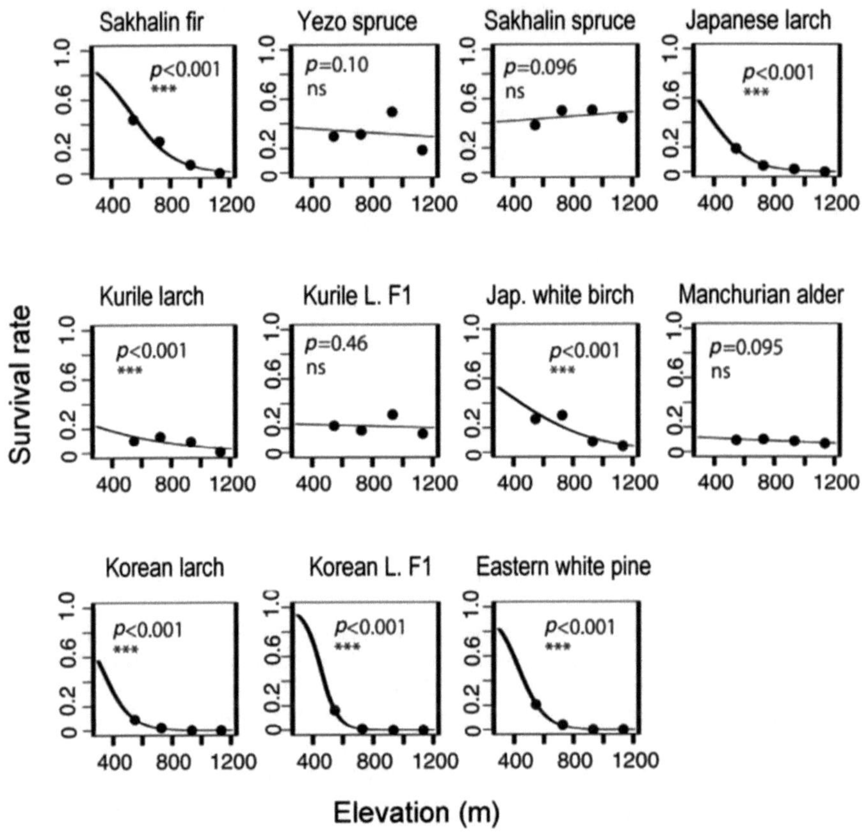

Fig. 10.15 Relationships between altitude and survival rate of 11 species in the elevation gradient experimental plantations in the UTHF. Kurile L. F1 and Korean L. F1 indicate F1 hybrid of Kurile larch × Japanese larch and F1 hybrid of Korean larch × Japanese larch, respectively. Asterisks indicate the statistical significance of the regression slope, tested by generalized linear models with binomial error. (Modified after Goto et al. 2014, with the publisher's permission)

Scots pine (*Pinus sylvestris*), Jack pine (*Pinus banksiana*), Japanese white birch, and Manchurian alder (*Alnus hirsuta* var. *hirsuta*) (Plot Nos. 3020–3023).

By 1979, Siberian larch, Scots pine, and Jack pine had almost completely died off, and American larch also experienced significant mortality. Among the remaining 11 species, the native spruce species (Yezo and Sakhalin spruce) and the F1 hybrid of Kurile larch × Japanese larch had higher survival rates at the high-elevation sites in 2008 (Goto et al. 2014, Fig. 10.15). These results indicated that Sakhalin spruce is the most promising species for planting in higher altitudinal zones (Goto et al. 2014).

In 1990, experimental plantations featuring native broadleaf species were established along an elevational gradient (Plot Nos. 3503–3506). The species included Japanese oak, Japanese elm (*Ulmus davidiana* var. *japonica*), Manchurian elm

(*Ulmus laciniata*), Japanese linden (*Tilia japonica*), Maximowicz's lime (*T. maximowicziana*), Japanese white birch, Erman's birch, Monarch birch (*B. maximowicziana*), Katsura tree (*Cercidiphyllum japonicum*), castor aralia (*Kalopanax septemlobus*), Manchurian ash (*Fraxinus mandshurica*), Japanese ash (*Fraxinus lanuginosa*), and Hokkaido bird cherry (*Padus ssiori*). These plantations were established at four elevation sites (400, 630, 780, and 860 m a.s.l.).

As of 2020, only a few or several individuals of Japanese oak, Japanese white birch, Erman's birch, and castor aralia survived at the highest elevation site (UTHF, unpublished data). Even at the second and third highest elevation sites, all species, except for these four, nearly disappeared. At the lowest elevation, several species, including Manchurian elm, Japanese ash, and Hokkaido bird cherry, had nearly vanished by the spring of 2013, while other species persisted. However, following significant snowpack damage in the autumn of 2013 (refer to Chap. 13), only a few species including Manchurian ash and the previously mentioned four species were thriving in 2020.

These findings suggest that planting broadleaf species is challenging, particularly at elevations greater than 600 m a.s.l. The four species, Japanese oak, Japanese white birch, Erman's birch, and castor aralia, may be relatively suitable for plantations in these conditions.

10.3 Improvement of Planting Materials

10.3.1 Introduction of Exotic Tree Species

The native coniferous species Sakhalin fir, Yezo spruce, and Sakhalin spruce are slow-growing. In the early stages of the SSMS, the "Clear-Cutting Forest," where natural regeneration was challenging, and future quantitative and qualitative growth could not be reasonably expected (refer to Chap. 8), were planned to be replaced with plantation forests. This forest type, despite having slightly high soil moisture, was generally fertile, making afforestation reliable. Furthermore, converting them into plantation forests was expected to increase forest productivity (Takahashi 2001). To achieve early recovery of growing stock, fast-growing exotic tree species suitable for Hokkaido were needed.

In the UTHF, exotic tree species were introduced from meteorologically similar regions of East Asia, North America, and Europe during 1905–1930, and an arboretum and a sample plantation site were established in 1931 and 1917, respectively. However, the origins or provenances of these materials were unclear. After that, exotic tree species with clear provenance records were reintroduced during 1955–1965 and two sample plantation sites were established by 1965. These materials from the second introduction were collected from multiple provenances in North American and northern European countries, focusing on species of pines (*Pinus*), spruces (*Picea*), firs (*Abies*), Douglas fir (*Pseudotsuga*), larches (*Larix*), birches (*Betula*), poplars (*Poplus*), and alders (*Alnus*).

The growth conditions of the first set of materials at approximately 40 years old and the initial growth conditions of the second materials at 10 years old or younger were reported in Hamaya et al. (1975) and Takahashi et al. (1974). Through these experimental plantings, the adaptability of exotic species to Hokkaido, especially to the UTHF, was examined. The findings were as follows: species and races from northeastern North America grew well, especially eastern white pine, but those from western North America were inferior to those from northeastern North America. Species and races from northern or central Europe grew better than those from southern Europe, and Norway spruce, Scots pine, and European birch (*Betula pendula*) showed good performance. For firs, poplars, and alders, there were no species better than the native species Sakhalin fir, Japanese poplar (*P. suaveolens*), and Manchurian alder (*A. hirsuta*).

10.3.2 Breeding of Hybrid Larch

Japanese larch, *L. kaempferi*, was introduced to Hokkaido around 1880 from central Honshu, Japan. It was expected to be a suitable planting material for cool regions like Hokkaido due to its fast-growing and cold-tolerant characteristics. However, in Hokkaido, bark-feeding by rodents, particularly the subspecies of grey red-backed vole, *Craseomys rufocanus bedfordiae*, and gnawing by hares caused severe damage. Additionally, shoot blight and Armillaria root rot diseases severely affected *L. kaempferi* plantations after the 1960s (refer to Chap. 14).

Although other larch species were inferior to *L. kaempferi* in terms of growth, their tolerance or resistance to some biological and meteorological disasters was greater than that of *L. kaempferi* in Hokkaido. Therefore, to increase the resistance of *L. kaempferi* to bark-feeding by voles, cross-breeding with other larch species was conducted. This cross-breeding approach followed the successful example of *L.* × *eurolepis*, a hybrid of *L. kaempferi* × *L. decidua* in Europe, whose resistance to Lachnellula canker was enhanced by interspecific hybridization.

The UTHF played a crucial role in advancing the cross-breeding of *L. kaempferi* in Hokkaido. To produce fast-growing, vole-resistant, and disease-resistant hybrids, crossing among *Larix* species, varieties or races around the world have been conducted since 1955, and cross-ability, morphological and phenological characteristics, growth, resistance to voles, hares and diseases, and wood quality have been examined. The details are summarized in Kurahashi (1988) and Hamaya and Kurahashi (1981). Through the series of examination, the F_1 hybrid of *L. gmelinii* var. *japonica* from Kurile × *L. kaempferi* was revealed to be the best combination of species with high growth ability and high adaptivity in Hokkaido, which means that it has a high general combining ability.

The lineage combination, which is the individual of each sex for hybridization that demonstrated the best specific combining ability, was selected thereafter. When selecting the *L. gmelinii* var. *japonica* lineage, the unique characteristic of fruiting annually from the juvenile stage was emphasized due to its usefulness for seed orchards. Consequently, a breed named "Toen No. 1" was registered and has been referred for evaluating the characteristics of subsequent hybrids developed by other organizations, although it has unfortunately never been used for plantations in the UTHF.

Furthermore, the method of top-budding in which buds are grafted onto shoots within the crowns of adult trees was developed. This method significantly shortens the period needed for flowering and allows for early and massive production of hybrid seeds, as well as quick replacement of clones in seed orchards. At present, the heterosis of hybrids observed in the juvenile stage is not clear, while the superior growth and adaptivity of *L. gmelinii* var. *japonica* × *L. kaempferi* hybrids persisted until they were 40–50 years old (Sakaue et al. 2017).

Backcrossing, F_2 hybridization, and triple crossing were also conducted for further improvement of the above F_1 hybrid. This aimed to address defects in the hybrid such as relatively high susceptibility to needle cast and Armillaria root rot originating from maternal *L. gmelinii* var. *japonica*, and to enhance growth. However, the usefulness of these generations has not been confirmed, although backcrosses and triple crosses could enhance disease resistance (Hamaya and Kurahashi 1981; Sakaue et al. 2018).

10.3.3 Individual Selection for Genetic Improvement

In the SSMS, the traits or phenotypes of individual trees are emphasized from the viewpoint of their economic value or availability for lumber. Thus, individuals with poor phenotypes, which were thought to be genetic, were actively removed from natural forests under the concept that natural forest management is the operation of controlling genes (Takahashi 2001). This concept originated from the "plus tree" and "reserved tree" selection conducted in the 1950s.

"Plus trees" are individuals with superior traits, such as high diameter and height growth, and are expected to become elite trees in the future. "Reserved trees" are individuals that are next to plus trees in terms of desired traits and that have rare and unique traits. The selection was conducted for both coniferous and broad-leaved trees in both natural forests and plantations, resulting in a total of 98 plus trees and 252 reserved trees being selected, including exotic tree species.

Seed orchards and clone stocks were established from this genetic reserve, contributing to the genetic improvement of natural forests in the UTHF (refer to Chap. 11) and to the above-mentioned breeding of hybrid larch.

References

Chazdon R, Calixto B, Oliveira M, Messinger J, Alves J, Calmon M, Anderson W (2022) The benefits and power of assisted natural regeneration. World Resource Institute. https://www.wri.org/insights/what-assisted-natural-regeneration-benefits-definition

Dobor L, Hlásny T, Zimová S (2020) Contrasting vulnerability of monospecific and species-diverse forests to wind and bark beetle disturbance: the role of management. Ecol Evol 10:12233–12245. https://doi.org/10.1002/ece3.6854

Fukuda E, Suyama T, Shibuya Y, Yagi T, Meguro R (2009) Influence of grazing cattle on survival and growth of Cryptomeria japonica D. Don seedlings in a Sasa kuriensis Makino et Shibata-predominated forest floor of a beech forest. Jpn J Grassl Sci 55:132–40

Goto S, Inukai H, Ogawa H, Suzuki S (2020) Current situation and issues of natural regeneration practice of *Betula maximowicziana* in secondary forests resulting from forest fires. For Tree Breed Hokkaido 63(2):15–19. (in Japanese)

Goto S, Tsuda S (2007) Scarification for promoting sustainable resource management of *Betula maximowicziana* in a second-growth forest. Jpn J For Soc 89:138–143

Goto S, Iijima H, Kimura N (2010) Long-term effects on tree regeneration of soil scarification with microtopography manipulation in mixed forests of Central Hokkaido, northern Japan. J For Res 15:328–336

Goto S, Owari T, Tokuni M, Matsui M (2014) Survival and growth of eleven tree species after the elapse of more than 45 years since planting at four different altitudinal zones in the University of Tokyo Hokkaido Forest. Bull Tokyo Univ For 130:15–29. (in Japanese with English summary)

Hamaya T, Kurahashi A (1981) Breeding of larch by species hybridization in Japan. Proc XVII IUFRO World Congress Div 2:157–168

Hamaya T, Kurahashi A, Sasaki T, Ogasawara S (1975) Lists and growth records of the woody plants cultivated in the arboretum and the sample plantations of the university Forest in Hokkaido. Misc Inform Tokyo Univ For 19:14–54. (in Japanese with English summary)

Harada A, Yoshida T, de Dios VR, Noguchi M, Kawahara T (2009) Growth and survival of tall tree seedlings 6-8 years after soil scarification in Hokkaido. J Jpn For Soc 90:397–403. (in Japanese with English summary)

Hatano K (1984) Tennen koushin mondai no saikin no ugoki [Recent developments in the issue of natural regeneration]. North For, Japan 36:187–190

Hirata M, Shishido K, Yamamoto H, Okamura K, Asama M, Takahashi Y, Kawahara S (1996) Growth characteristics of *Betula platyphylla* var *japonica* in a density-experiment plot. Trans Mtg Hokkaido Br Jpn For Soc 44:83–85. (in Japanese)

Ishibashi S, Takahashi M, Sasaki S (2010) Growth analysis of *Abies sachalinensis*, *Picea jezoensis* and *Piccea glehnii* which were planted in natural forest. Trans Mtg Hokkaido Br Jpn For Soc 58:105–106. (in Japanese)

Isozaki Y, Owari T, Koike Y, Fukuoka S, Oikawa N (2008) Growth progress of a 100-year-old planted larch stand in the Tokyo University Forest in Hokkaido. Trans Mtg Hokkaido Br Jpn For Soc 56:117–119

Ito E, Hashimoto T, Aizawa S, Ishibashi S (2018) Contemporary significance of soil scarification for promoting tree regeneration in Hokkaido, northern Japan. Jap J For Environ 60:71–82. (in Japanese with English summary)

Iwamoto K, Nishida N (1963) Shokusai mitsudo wo inishita shirakanba no seicyou siryou [Growth records of Japanese white birch plantation with different stand densities (Translated by authors from Japanese title)]. Trans Mtg Hokkaido Br Jpn For Soc 12:57–60. (in Japanese)

Iwamoto K, Nishida N (1968) Karamatsu-Todomatsu konshoku no seiseki to mondaiten (Results and problems of Saghalin fir-Japanese larch mixed plantation). Trans Mtg Hokkaido Br Jpn For Soc 17:53–56. (in Japanese)

Iwamoto K, Nishida N, Takada K (1970) Shokusai mitsudo wo inishita shirakanba no shoki seicyou (Initial growth of Japanese white birch plantation with different stand densities). Trans Mtg Hokkaido Br Jpn For Soc 19:65–67. (in Japanese)

Karthigesu J, Owari T, Tsuyuki S, Hiroshima T (2023) UAV photogrammetry for estimating stand parameters of an old Japanese larch plantation using different filtering methods at two flight altitudes. Sensors 23:9907. https://doi.org/10.3390/s23249907

Kurahashi A (1988) Improvement of larches by species hybridization. Bull Tokyo Univ For 79:1–94. (in Japanese with English summary)

Kurahashi A, Takahashi Y, Hirokawa T, Sato I, Kasahara H (2001) Karamatsu 93-nensei jinnkourin no seicyou sozaitokusei –Toudai Hokkaido Ensyurin 1908 shokusai—[Growth and wood characteristics of 93-year larch plantation: planted in 1908 in the University of Tokyo Hokkaido Forest (Translated by authors from Japanese title)]. For Tree Breed Hokkaido 44(1):20–23. (in Japanese)

Liu CLC, Kuchma O, Krutovsky KV (2018) Mixed-species versus monocultures in plantation forestry: development, benefits, ecosystem services and perspectives for the future. Glob Ecol Conserv 15:e00419

Matsui M, Okamura K (2004) Todomatsu to sutorobumatsu no mitsudobetsu sikenchi ni okeru zourinnboku to tennen koushinboku no seiiku joukyou [Growth situation of planted and naturally regenerated trees in stand density control experiments of *Abies sachalinensis* and *Pinus strobus* (Translated by authors from Japanese title)]. For Tree Breed Hokkaido 47(2):21–24. (in Japanese)

Nakagawa M, Takahashi Y, Watanabe S, Suzuki K (1994) Dynamics of seedlings which naturally regenerated after a horizon removal in boreal mixed forest. Trans Jpn For Soc 105:385–387. (in Japanese)

Owari T, Eguchi Y, Takuma R, Okamura K, Fukuoka S, Kimura N, Tatsumi S (2015) Development of a precision planting technique for securing tree regeneration in a natural forest managed by selection system: a preliminary trial. Boreal For Res 63:81–84. (in Japanese)

Sakaue D, Kimura N, Fukuoka S (2017) Growth of middle-aged hybrid larches in the University of Tokyo Hokkaido Forest: some cases of experimental plots which passed more than 50 years. For Tree Breed Hokkaido 60(2):14–19. (in Japanese)

Sakaue D, Kimura N, Fukuoka S (2018) Growth of middle-aged hybrid larches in the University of Tokyo Hokkaido Forest: some cases on F_2 and B_1 families between Japanese and Kurile or Prince Rupprecht's larches. For Tree Breed Hokkaido 61(2):1–7. (in Japanese)

Shibata Z (1981) Selection cutting system in the experiment forest of Hokkaido, Tokyo University. In: Oomae E (ed) Nihon no Takubatsu [Selection cutting system in Japan (Translated by authors from Japanese title)]. Japan Forestry Inventory Committie, Tokyo

Takahashi N (1951) Rinbun-Segyouhou Gairon – Ensyurin ni okeru rekishi to jikken—[Stand-based Silvicultural management system – experiments and practices in a university forest (translated by authors)]. Northern For Jpn 145–146:1–5. (in Japanese)

Takahashi N (1952) Eradicative process of Kumai-Sasa (*Sasa paniculate* Makino et Shibata) by over-grazing in the forest. Misc Inform Tokyo Univ For 9:7–16

Takahashi N (2001) Rinbun segyo ho: sono kangae to jissen (kaitei ban) [The Stand-based Silvicultural Management System: its Theory and Practices (Revised Ed.)]. Log Bee, Sapporo, 125pp (in Japanese)

Takahashi N, Hamaya T, Kurahashi A (1974) Records of the growth of young sample stands of introduced species, in the Tokyo University Forest in Hokkaido. Misc Inform Tokyo Univ For 18:29–66. (in Japanese with English summary)

Takahashi K, Suzuki SN, Goto S (2023) Conversion of a *Pinus strobus* plantation to a conifer-broadleaved mixed forest by small-scale clear-cutting and scarification treatments: regeneration dynamics of *P. strobus* and native tree species during 15 years. J Jpn For Soc 105:65–75

Tanimura R, Yamasaki K (2020) Jyusyu no tayousei wo hakaru kouzhin-houhou no Kaihatsu: jigaki negaesi kara 10 nen-go no seika to konngo no tenkai [Development of techniques for regeneration of diverse species: Results after 10 years from scarification and uprooting and future perspectives], Technical Exchange Symposium for Northern Land and Forest Development 2020. https://www.rinya.maff.go.jp/hokkaido/gizyutu_c/Presentation/attach/pdf/index-60.pdf. Last accessed Jan 15, 2024

The University of Tokyo Forests, Graduate School of Agricultural and Life Sciences, The University of Tokyo (2022) Education and research plan (2021-2030) of the University of Tokyo Forests Part 3 regional Forest plans the University of Tokyo Hokkaido Forest (the 14th period). Misc Inform Tokyo Univ For 64:103–190

Yamamoto H (1992) Studies on growth prediction of two-storied artificial stands (I) growth characteristics of mixed stands of Japanese larch and Sakhalin fir. Bull Tokyo Univ For 88:33–57. (in Japanese with English summary)

Yamamoto H (1993) Studies on growth prediction of two-storied artificial stands (II) growth model of mixed planting two-storied stands of Japanese larch and Sakhalin fir. Jpn J For Res 75:65–69. (in Japanese)

Yasue K (2000) Nikuyou-gyu ni yoru shinrin riyou no kanousei - 21 seiki ni muketa rinkan houboku no houkousei to kadai - [Possibilities of Forest Utilisation with Beef Cattle: Directions and Challenges of Silvopastoral Grazing toward the 21st Century]. Res Bull Livest Farm Fac Agric Hokkaido Univ 17: 67–76

Yoshida T, Iga Y, Ozawa M, Noguchi M, Shibata H (2005) Factors influencing early vegetation establishment following soil scarification in a mixed forest in northern Japan. Can J For Res 35(1):175–188. https://doi.org/10.1139/x04-156

Chapter 11
Seedling Production

Daisuke Sakaue, Susumu Goto, and Satoshi N. Suzuki

Abstract Under the Stand-based Silvicultural Management System, conifer seedlings are planted in natural forests to assist natural regeneration. For this purpose, seedlings of *Abies sachalinensis*, *Picea jezoensis*, and *Picea glehnii*, which are major species of natural forest, are raised in a nursery at the University of Tokyo Hokkaido Forest (UTHF). Clonal seed orchards have been established as seed sources for *A. sachalinensis* and *P. jezoensis*, while seed stands are designated in natural forests for *P. glehnii*. To minimize genetic disturbance and maintain genetic diversity, ortets of seed trees that constitute the seed orchards were selected from natural forests within the UTHF. The seedlings are outplanted after a raising period of 6–7 years. Most of the raised seedlings are bare-rooted, while containerized seedlings are currently under trial. During the raising period, some diseases, such as *Racodium* snow blight and damping-off, threaten seedling production. These diseases are managed through cultural and chemical control methods. Care should also be taken to mitigate damage caused by extreme climatic events such as late frost, heat waves, and drought under ongoing climate change.

Keywords Nursery · Seed orchard · Genetic disturbance · Genetic diversity · Disease control · Bare-rooted seedling · Containerized seedling

D. Sakaue (✉)
The University of Tokyo Chichibu Forest, Graduate School of Agricultural and Life Sciences, The University of Tokyo, Chichibu, Japan
e-mail: sakaue@uf.a.u-tokyo.ac.jp

S. Goto
The University of Tokyo Tanashi Forest, Graduate School of Agricultural and Life Sciences, The University of Tokyo, Nishitokyo, Japan
e-mail: gotos@uf.a.u-tokyo.ac.jp

S. N. Suzuki
The University of Tokyo Hokkaido Forest, Graduate School of Agricultural and Life Sciences, The University of Tokyo, Furano, Japan

Nakagawa Experimental Forest, Field Science Center for Northern Biosphere, Hokkaido University, Otoineppu, Japan
e-mail: snsuzuki@fsc.hokudai.ac.jp

T. Owari et al. (eds.), *Integrative Forest Management and Silviculture*, Ecological Research Monographs, https://doi.org/10.1007/978-981-95-1017-7_11

11.1 Introduction

At the University of Tokyo Hokkaido Forest (UTHF), seedlings are planted to assist the natural regeneration of native conifers under the Stand-based Silvicultural Management System (refer to Chap. 10). Exceptions are made for larch plantations in the plantation working group, where purchased seedlings of introduced tree species are planted as monocultures. Therefore, the target species for seedling production are limited to the native species, including *Abies sachalinensis* (Sakhalin fir, abbreviated as *As* hereafter), *Picea jezoensis* (Yezo spruce, abbreviated as *Pj*), and *Picea glehnii* (Sakhalin spruce, abbreviated as *Pg*). These species are not only dominant in the natural forests of the UTHF but also key species that characterize the hemiboreal conifer-broadleaf mixed forest in this region (refer to Chap. 5).

Seeds for raising these seedlings are collected within the UTHF to minimize disturbances to species and/or the genetic composition of the natural forests. This principle was established in 1996, although seedlings from outside the UTHF were purchased to compensate for shortages during the restoration following the typhoon damage in 1981. Additionally, seedlings of exotic species were planted during restoration after the forest fires in the 1910s and the typhoon damage in 1954 (refer to Chap. 13).

Seedlings have been produced in the nursery of the UTHF since its establishment in 1906 (Fig. 11.1). This nursery covers an area of 3.82 ha and is located on a river terrace of the Sorachi River, with lowland soil derived from river sediments. A total

Fig. 11.1 Nursery of the University of Tokyo Hokkaido Forest. (Photo by Daisuke Sakaue, Octoer 18, 2016)

of 18,000 seedlings (7400 for *As*, 7900 for *Pj*, and 2800 for *Pg*, on average) are produced annually for outplanting, with 17,000 being bare-rooted and 1000 containerized seedlings (The University of Tokyo Hokkaido Forest 2022). This chapter describes the characteristics and the outline of seedling production at the UTHF.

11.2 Seed Sources and Seed Preparation

When collecting seeds, minimizing genetic disturbance and maintaining genetic diversity are emphasized. Therefore, it is crucial to use proper seed sources for each species. Cones with open-pollinated seeds of *As* are collected from a clonal seed orchard (Fig. 11.2) established in the 72A subcompartment of the UTHF in 1960. This seed orchard covers an area of 1.0 ha and consists of 16 clones. The ortets of these clones were plus trees selected from natural forests in the UTHF (refer to Chap. 10) that were significantly superior to surrounding trees in height, diameter, straightness, and branch height. Ramets of these plus trees were propagated by grafting during the period 1956–1958. A total of 958 ramets, with 33–70 ramets per clone, were planted at 3-m intervals in a systematic square arrangement. No pruning has been conducted. The seed orchard is surrounded by farmlands and is approximately 1 km away from the forest, minimizing pollen contamination. Abundant

Fig. 11.2 Clonal seed orchard of *Abies sachalinensis*. (Photo by Susumu Goto, March 4, 2024)

seeds were first obtained in 1974. Cones with open-pollinated seeds have been collected by felling seed trees in mass-fruiting years since 1985, and thinning seed trees has been conducted simultaneously. Recently, abundant fruiting has not been observed, which is speculated to be due to disturbed pollen flow caused by the high density of seed trees (Sakaue et al. 2019). Therefore, the stem density was reduced to 205 stems/ha by thinning in 2022. Further investigations into pollen flow within seed orchards and from external sources are necessary to evaluate the productivity and quality of seeds. Additionally, understanding the differences in progeny characteristics among maternal clones and the adaptivity of progenies at planted sites is needed to improve the seed orchard.

Cones with open-pollinated seeds of *Pj* are collected from naturally growing healthy trees that are felled in areas where logging operations are conducted during years of abundant fruiting. This method, adopted since the 2000s, is labor-intensive, because the felled trees from which cones are collected are scattered throughout natural forests, and the collection sites are far from the nursery. Moreover, the genetic characteristics of seed trees vary among harvesting years, and excellent seed trees are lost once seeds are collected. There is also concern about the decreasing number of suitable seed trees due to the recent decrease in *Pj* in natural forests. In the past, a seed stand was designated and cones were collected by climbing seed trees. However, the seed stand was abolished after severe damage by the typhoon in 1981. Consequently, a seed orchard was newly established in 2017 (Sakaue et al. 2018, Fig. 11.3). This clonal seed orchard covers a total area of 0.73 ha, with 201 ramets from 60 clones planted at 5-m intervals (400 trees/ha). The ortets of these

Fig. 11.3 Clonal seed orchard of *Picea jezoensis*. (Photo by Daisuke Sakaue, August 25, 2022)

clones were selected from naturally growing trees throughout the major area of *Pj* distribution in the UTHF based on their superior characteristics, such as diameter, straightness, taper, and branch height compared to surrounding trees. The plus trees selected in the 1950s and 1960s were not used for ortets, because their branch height was too high to collect scions for grafting. Ramets were propagated by grafting during the period 2010–2014 in cooperation with the Hokkaido Regional Breeding Office, Forest Tree Breeding Center, Forest Research and Management Organization, Japan. The planting design was primarily based on a random arrangement, but adjustments were made to ensure that individuals of the same clone were not adjacent to each other and that no single individual was adjacent to multiple individuals of the same clone. To increase the number of clones as much as possible, the number of ramets of each clone was limited to a maximum of four. There are few *Pj* trees around the seed orchard, minimizing pollen contamination. Flowering has already been observed in several ramets since 2017, and seed harvesting is expected in the near future. This is the first establishment of a seed orchard for *Pj* in Japan.

Cones of *Pg* were previously collected from a designated seed stand in natural forests in the subcompartment 8A by climbing seed trees after open pollination (Fig. 11.4). In addition to this stand, cones of *Pg* were sometimes collected from *Pg* stands in the subcompartments 9B and 13B. Recently, the designated seed stand has declined and the designation has been canceled. New seed stands were designated in the subcompartments 10B and 27B in 2020. Seed stands require abundance of mature trees, but *Pg* grows sparsely in most natural forests, and *Pg*-dominant forests

Fig. 11.4 Seed stand of *Picea glehnii*. (Photo by the UTHF, July 16, 2021, licensed under CC BY-NC-ND 4.0)

are limited to specific sites such as moors, rocky areas, lapilli, serpentine, and higher elevations. Because populations established in these habitats are isolated, the population size sufficient for random mating and the genetic diversity of the population should be examined when selecting seed stands (Goto et al. 2009, 2011).

For each species, cones are collected from late August to mid-September in years of mass fruiting. The collected cones are sun-dried, and winglets are mechanically removed from the obtained seeds. After the contaminants are carefully removed, the seeds are stored at −5 °C. Seeds can be stored for more than 8 years (*Pj* and *Pg*) or 11 years (*As*) without losing their germination ability.

11.3 Nursery Practices for Bare-Root Seedling Production

In Hokkaido, the production of *As* and *Pg* seedlings has been widely conducted at the commercial level. In contrast, seedling production of *Pj* has been inactive, because it has been avoided in commercial plantations. The reasons include nursery diseases (refer to Sect. 11.5 for details), damage to young plantations, such as late frost injuries and bud galls caused by aphids (*Adelges japonicus*), and the fundamentally slow growth of *Pj* (Koshika and Seino 1966). In the UTHF, seedlings of *Pj* are raised stably using the following method, in addition to *As* and *Pg*.

The seedlings of the three species are raised using the same method, with bare-root seedling production taking six years. After sowing, germinated seedlings are kept for 2 years on seedbeds (Fig. 11.5) and then they are transplanted to new beds (seedling beds) and grown for 4 or 5 years (Fig. 11.6). As a result, 6- or 7-year-old seedlings are outplanted. The growth period of *As* and *Pg* in the UTHF nursery is longer than other forest nurseries in Hokkaido, where the process typically takes 4–6 years (2 years in seedbeds and 2–4 years in seedling beds).

Seeds are sown in seedbeds in late autumn (mid-October) to avoid labor concentration in spring, although they can also be sown in early spring. The number of seeds sown depends on the maturation rate of the seed lot. The expected density after germination in seedbeds is several thousand seedlings per square meter, which are then thinned to approximately 1000 seedlings/m^2 in the second spring. Seedlings are later transplanted at a density of 36 seedlings/m^2 to the seedling beds. During the third and fourth years, and if necessary, the second year as well, both lateral and longitudinal roots are pruned by a root-cutting plow attached with a tractor at seedling beds. Root pruning prevents excessive elongation of seedlings and encourages the emergence of fine roots. As a result, raised 6- or 7-year-old seedlings are outplanted from September to December during the final year of the raising period (fourth or fifth year in the seedling beds). Further details of these practices are described in Ogasawara (2001).

Fig. 11.5 Seed bed of *Picea jezoensis* maintained for two growing seasons. (Photo by Daisuke Sakaue, October 18, 2016)

11.4 Production of Containerized Seedlings

Recently, containerized seedling production has become common in Japan, including Hokkaido (Forestry Agency 2023). Containerized seedlings are grown in air-pruning cells without root looping, resulting in steady root balls. This characteristic enhances survival rates compared with those of bare-rooted seedlings by preventing root injury during outplanting and extends the suitable period for outplanting. Moreover, containerized seedlings require smaller planting holes and eliminate the need to backfill roots. Additionally, since containerized seedlings are portable, the timing of their bud flushing can be managed by storing them at cold temperatures, such as in a large refrigerator. This allows outplanted seedlings to avoid frost damage at planting sites, which are usually cooler than nurseries. Compared with bare-root seedling production, the production of containerized seedlings can save labor for weeding because few weed seeds are contained in the culture media.

The UTHF started to raise containerized seedlings of three native conifers, *As*, *Pg*, and *Pj*, in 2009 (Fig. 11.7). The production of containerized seedlings had already started at that time in Hokkaido, but no standardized systems were available for these three conifer species. The UTHF subsequently developed methods by collaborating with the Forestry and Forest Products Research Institute, Japan, and the

Fig. 11.6 Seedling bed of *Picea jezoensis* just before outplanting. (Photo by Daisuke Sakaue, October 18, 2016)

Forestry Research Institute of the Hokkaido Research Organization. As a result, the collaborating group created a manual for containerized seedling production for native conifers ("Short-term and healthy *Picea jezoensis* seedling production" Research Group 2014).

For all three species, seedlings grown in seedbeds for two years are transplanted to container cells with the capacity of 150 mL or 300 mL. In countries such as Sweden, which are advanced in the use of containerized seedlings, direct seeding into containers is common because of the high germination rate (Okada et al. 2012). In contrast, the accuracy of discriminating viable and nonfilled seeds is low in Japanese tree species for plantations. Therefore, seeds are sown in seedbeds to avoid containers containing many blank cells. After two years of growth in containers in a greenhouse, the seedlings can be used for outplanting ("Short-term and healthy *Picea jezoensis* seedling production" Research Group 2014). The cultivation conditions of the containers, such as fertilization, irrigation, and day length, were examined (Ogawa et al. 2013).

However, the growth of 3- or 4-year-old containerized seedlings (1 or 2 years in containers after transplanting) fell short of that of 7-year-old bare-root seedlings in early trials of outplanting due to their shorter initial heights (Owari et al. 2013; Goto et al. 2015; Kimura et al. 2023), although the survival rate of 4-year-old

Fig. 11.7 (**a**) Production of containerized *Picea jezoensis* seedlings (Photo by Susumu Goto, October 12, 2011). (**b**) Containerized *P. jezoensis* seedlings (300 mL) for outplanting (Photo by Susumu Goto, October 2, 2010)

containerized seedlings was comparable to that of 7-year-old bare-root seedlings. Consequently, 6-year-old containerized seedlings (4 years after transplanting to the container) are raised in the UTHF. Containerized seedlings were initially expected to have the potential to shorten the raising period via an intensive culture system. Since a longer raising period increases the risk of disease, pest, and climatic damage, further research is desirable to shorten the raising period.

One advantage of using containerized seedlings is their ability to be planted throughout the year, offering a solution for optimizing nursery operations. To facilitate this, Goto et al. (2013) attempted to store containerized seedlings in a refrigerator. In their study, they compared the survival rates between storage boxes made of standard and those made of Controlled Temperature Method (CTM) cardboard. CTM cardboard boxes have been shown to be effective for refrigerator storage (Fig. 11.8).

In the early 2000s, when the UTHF examined culture methods, the use of containers without side slits was common in Japan. However, in countries such as

Fig. 11.8 Survival of containerized *Picea jezoensis* seedlings after storage in a refrigerator. The survival rate was significantly greater for seedlings stored in a CTM cardboard box (left) than for those stored in a standard cardboard box (right). (Photo by Susumu Goto, October 2, 2010)

Sweden, containers with side slits were commonly used, because they improved rooting systems. At the UTHF, the use of side-slit containers has proven to be more effective than the use of nonslit containers in enhancing the survival rate of *As* seedlings during the 3-year raising period (Fukuoka et al. 2022), potentially due to the reduced moisture level in the containers. Diseases such as gray mold are frequently observed in *As* seedlings grown in nonslit containers. Conversely, the survival rates of *Pj* and *Pg* seedlings were slightly higher in nonslit containers than in those with side slits. Although the exact reason for this phenomenon remains unclear, the relatively high moisture content within nonslit containers may be advantageous for *Picea* species.

11.5 Nursery Diseases and Their Control

The major diseases in the nursery are *Racodium* snow blight[1] and damping-off (Figs. 11.9 and 11.10, respectively). Additionally, gray mold sometimes occurs severely (Fig. 11.11). For containerized seedlings, there are no conspicuous diseases thus far. Care should be taken against gray mold, because containerized seedlings tend to be overcrowded, resulting in a humid environment in which the disease easily occurs.

Pj and *Pg* seedlings are often severely damaged by *Racodium* snow blight, not only on forest floors but also in nurseries, whereas *As* seedlings are less affected than *Pj* and *Pg* seedlings. Small seedlings in seedbeds can be killed, because they bend down under snow cover, causing their bodies to be pushed against the ground surface where pathogenic fungi are present. In contrast, damage to larger seedlings older than 5 years is limited to the death of lower branches, because they don't bend down, except for their lower branches. Therefore, it is important to raise seedlings

[1] The outline of *Racodium* snow blight and the taxonomical situation of its pathogen are described in Chap. 14.

Fig. 11.9 *Racodium* snow blight on 2-year-old *Picea jezoensis* seedlings at the seedbed. (Photo by Daisuke Sakaue, April 10, 2009)

Fig. 11.10 Damping-off of current-year-old *Picea jezoensis* seedlings at the seedbed. (Photo by Daisuke Sakaue, May 20, 2011)

Fig. 11.11 Gray mold on 6-year-old *Abies sachalinensis* (left) and *Picea jezoensis* (right) seedlings. (Photo by Daisuke Sakaue, July 14, 2014)

Fig. 11.12 Snow sheds installed over seedbeds. (Photo by Daisuke Sakaue, March 5, 2009)

with larger diameters that are less likely to bend down. The density of seedlings in the seedbeds is also important, because sparsely growing seedlings are more likely to bend under snow pressure. To reduce snow pressure and prevent seedlings from bending, snow sheds are installed over seedbeds during winter to minimize snow accumulation (Fig. 11.12). Additionally, the fungicide oxine-copper (copper 8-hydroxyquinoline, 8-quinolinol) is sprayed before the seedlings are covered with snow in autumn. After germination, the surface of the seedbeds is covered by lapilli, an inorganic volcanic product, to prevent seedlings from being covered with soil displaced by raindrops (Fig. 11.13). This approach also helps prevent seedlings from coming into contact with pathogenic fungi in the soil. In late March, snowmelt agents are scattered to shorten the favorable period for the pathogen. These countermeasures have enabled the successful production of *Pj* seedlings by mitigating the fatal impact of *Racodium* snow blight.

Damping-off is caused by several soil-borne pathogens, including *Fusarium* spp., for example, *F. oxysporum*, and *Rhizoctonia solani*. In addition to the above pathogens, *Calonectria morganii* (= *Cylindrocladium scoparium*) also affects *As* seedlings, while *Globisporangium debaryanum* (≡ *Pythium debaryanum*) can impact *Pj* and *Pg* seedlings (The Phytopathological Society of Japan 2024). The occurrence of damping-off is limited to current-year seedlings that have just germinated and are unlignified; however, root rot caused by *Fusarium* spp. can affect older *As*, *Pj*, and *Pg* seedlings. The symptoms of damping-off typically manifest as rot near the ground level, but sometimes include top rot. Damage caused by *Rhizoctonia* occurs in wet conditions when snowmelt water is still present; thereafter, damage caused by *Fusarium* spp. occurs when the seedbeds dry. Compared with *As* seedlings, *Pj* and *Pg* seedlings experience more severe damage. Inoculation tests with *Rhizoctonia solani* demonstrated that *Pj* seedlings are highly susceptible, *Pg* seedlings are moderately susceptible, and *As* seedlings are relatively resistant, with tolerance increasing with age in weeks. The anastomosis group of *Rhizoctonia solani* isolated from *Pj* and *Pg* seedlings in the UTHF nursery was identified as

Fig. 11.13 Lapilli scattered on seedbeds. (Photo by Daisuke Sakaue, October 1, 2009)

Fig. 11.14 Seeds of *Abies sachalinensis* coated with thiuram (looks blue in the photo), just before being covered with soil after sowing. (Photo by Daisuke Sakaue, October 23, 2008)

AG-5 (Sakaue, unpublished data). To control damping-off, seeds are coated with the thiuram fungicide before sowing (Fig. 11.14). Fungicides, such as hydroxyisoxazole (hymexazole), tolclofos-methyl, or flutolanil, are sprayed at the onset of disease occurrence. In the past, bed soils were burned for sterilization against soil-borne pathogenic fungi and to kill weed seeds; however, their effectiveness was unclear, because only the surface of the beds could be burned.

Gray mold is caused by *Botrytis cinerea*, and its symptom is bud or shoot blight. It occurs on old-growth seedlings of *As* and *Pj*, especially on *As*. Since the humid environment created by overcrowded seedlings enhances the occurrence of gray mold, making it important to maintain a lower seedling density, particularly in the second year in seedbeds.

In forest nurseries in Hokkaido, additional seedling diseases have been reported, including seed rot of *As* caused by *Sclerotium* sp., black root rot of *As* caused by *Macrophomina phaseolina*, web-blight of *As* caused by *Rhizoctonia solani* (=

Thanatephorus cucumeris), Rosellinia needle blight of *As* and *Pj* caused by *Rosellinia herpotrichioides*, and root lesions of *As*, *Pj* and *Pg* caused by *Pratylenchus penetrans* (The Phytopathological Society of Japan 2024). However, these diseases currently seldom occur in the nursery of UTHF.

11.6 Weather and Climate-Related Damages

Seedling losses caused by weather and climatic conditions are also a significant concern in nurseries. Late frost injury and winter desiccation can threaten seedlings (Fig. 11.15). Severe late frost damage was observed in *Pj* and *As* in 2008 (Ogawa et al. 2010). With climate change, the timing of bud flush has gradually advanced,

Fig. 11.15 (**a**) Late frost injury of *Abies sachalinensis* seedlings (photo by the UTHF, May, 2008, licensed under CC BY-NC-ND 4.0) and (**b**) winter desiccation of 2-year-old *Picea glehnii* seedlings at the seedbed (Photo by Daisuke Sakaue, April 10, 2009)

increasing the risk of damage from late frosts due to early bud flush, even though it prolongs the growing period. Additionally, frost heaving, which uproots juvenile seedlings in late autumn, often leads to seedlings mortality in the following spring. Proper drainage of beds is crucial for mitigating frost heaving. Notably, *Pj* is particularly vulnerable to late frost injury and frost heaving.

Heatwaves and droughts have also become increasingly problematic due to climate change. The mortality of 1-year-old seedlings, particularly *Pj* and *Pg*, from seedbeds has occurred several times in the spring over the past 15 years (Sakaue 2017). The cause of this mortality remains unknown, but it is possible that the damage is related to heat canker caused by excessively high soil surface temperatures. During the summer of 2021, when precipitation in July totaled only 16.0 mm—far below the July average of 140 mm— approximately half of the seedlings that had just been transplanted to the seedling beds in the spring died (Tanaka et al. 2023). With predictions of increased severity and frequency of such extreme climatic events due to ongoing climate change, these issues must be addressed as significant concerns in the near future (Figs. 11.16 and 11.17).

Fig. 11.16 (**a**) Mortality of 1-year-old *Picea jezoensis* seedlings at the seedbed and (**b**) the lesion suspected to be heat canker. (Photo by Daisuke Sakaue, May 17, 2017)

Fig. 11.17 Severe damage to transplanted seedlings (3-year-old) in seedling beds during drought years. (Photo by Satoshi Fukuoka, August 14, 2021, used with permission from the UTHF, licensed under CC BY-NC-ND 4.0)

References

"Short-term and healthy *Picea jezoensis* seedling production" Research Group (2014) Manual for short-term and healthy *Picea jezoensis* seedling production [Translated by authors from Japanese author name and title]. https://repository.dl.itc.u-tokyo.ac.jp/records/43077 (in Japanese)

Forestry Agency (2023) Annual report on forest and forestry in Japan, fiscal year 2022, Forestry agency, Ministry of Agriculture, Forestry and Fishery, Japan

Fukuoka S, Kimura N, Matsui M, Suzuki SN (2022) Raising of container seedlings in the University of Tokyo Hokkaido Forest: comparison of 3 native conifers and effects of side-slits of container. For Tree Breed Hokkaido 65(2):15–20. (in Japanese)

Goto S, Tsuda Y, Koike Y, Lian C, Ide Y (2009) Effects of landscape and demographic history on genetic variation in *Picea glehnii* at the regional scale. Ecol Res 24:1267–1277

Goto S, Ogawa H, Ishizuka W, Matsui M, Kimura N, Okamura K (2011) Effects of seed sources in university Forest in Hokkaido, the University of Tokyo, on survival and growth of planted trees in *Picea glehnii*. Bull Tokyo Univ For 124:107–118. (in Japanese with English summary)

Goto S, Ogawa H, Kimura N, Fukuoka S, Owari T (2013) Storage of containerized *Picea jezoensis* seedlings using a freezing chamber during winter season. For Tree Breed Hokkaido 65(2):27–30. (in Japanese)

Goto S, Owari T, Ogawa H, Kimura N, Fukuoka S, Takuma R, Inukai S, Takahashi K, Sasaki S (2015) Survival and shoot growth of containerized and bare-rooted seedlings of *Picea jezoensis* during three years after planting. For Genetics Tree Breeding 4:1–6. (in Japanese with English summary)

Kimura N, Suzuki SN, Fukuoka S, Matsui M, Goto S, Owari T (2023) Growth and survival of containerized Yezo spruce seedlings after 10-years from planting in the University of Tokyo Hokkaido Forest. For Tree Breed Hokkaido 64(2):29–33. (in Japanese)
Koshika K, Seino M (1966) A study on Ezomatsu (*Picea jezoensis*) resources in Hokkaido (II) history of Ezomatsu planting and its future challenges. Jpn J For Plann 26:73–84. (in Japanese with English summary)
Ogasawara S (2001) Nursery practices concerning with propagation of *Picea jezoensis* from seedlings at the Tokyo university Forest in Hokkaido. Bull Tokyo Univ For 106:49–68. (in Japanese with English summary)
Ogawa H, Takahashi K, Inukai S, Oya K, Kimura N (2010) Late frost damages occurred in 2008 at the Tokyo university Forest nursery in Hokkaido. For Tree Breed Hokkaido 58:131–134
Ogawa H, Kimura N, Fukuoka S, Goto S (2013) How do the form and volume of a container affect the survival and growth of containerized seedlings? For Tree Breed Hokkaido 65(2):17–18. (in Japanese)
Okada M, Ogawa H, Ishizuka W (2012) Sweden ni okeru shinyoujyu kontena-nae seisann no jirei syoukai [a case study of production of conifer containerized seedling production in Sweden] Northern For Japan 64(4):101–104. (in Japanese)
Owari T, Kimura N, Ogawa H, Fukuoka S, Takuma R, Inukai S, Takahashi K, Sasaki S, Goto S (2013) Plantation trials of containerized *Picea jezoensis* seedlings at the University of Tokyo Hokkaido Forest. For Tree Breed Hokkaido 56(2):35–36. (in Japanese)
Sakaue D (2017) Damping-off of 1-year old *Picea jezoensis* and *P. Glehnii* seedlings possibly caused by a snow blight fungus, *Racodium therryanum*, in non-snow covering summer season. Boreal Forest Res 65:59–62. (in Japanese)
Sakaue D, Kimura N, Fukuoka S, Goto S (2018) Establishment of the seed orchard for *Picea jezoensis* in the University of Tokyo Hokkaido Forest. For Tree Breed Hokkaido 61(1):15–22. (in Japanese)
Sakaue D, Kimura N, Fukuoka S, Goto S (2019) Initial growth differences of *Abies sachalinensis* seedlings among mother tree families comprising a seed orchard. Boreal For Res 67:29–32. (in Japanese)
Tanaka N, Ogawa H, Fukuoka S, Goto S (2023) 2021-nen ni Tokyo daigaku Hokkaido ensyurin no naehata de hassei sita kansougai ni tsuite [drought damages of conifer seedlings in the nursery of the University of Tokyo Hokkaido Forest in 2021]. Forest Gen Tree Breeding 12(3):108–110. (in Japanese)
The Phytopathological Society of Japan (2024) Common names of plant diseases in Japan (2024.2 edition). (in Japanese)
The University of Tokyo Forests (2022) Education and research plan (2021–2030) of the University of Tokyo Forests, Part 3 Regional forest plans, The University of Tokyo Hokkaido Forest (The 14th period). Miscellaneous Information of The University of Tokyo Forests 64:103–190. (in Japanese). doi:https://doi.org/10.15083/0002002908.

Part IV
Management and Conservation of Multiple Ecosystem Functions

Chapter 12
Effects of Forest Management on Ecosystem Processes

Satoshi N. Suzuki

Abstract Large-scale practices, such as clearcutting and plantations, can largely impact ecosystem processes. Specifically, it can reduce the amount of dead wood, which is essential for tree regeneration and organisms involved in saproxylic food webs. Although selection cutting can also reduce the amount of dead wood even under the SSMS, compared with unmanaged forests, the amount of dead wood in the UTHF is thought to be sufficient for sustaining the biodiversity of dead-wood-dependent species. Tree can have a large impact on forest floor conditions, especially by destroying tree saplings. It can also affect the dynamics of bark beetles. If tree harvesting leads to the creation of more newly clear-cut sites or sparse forests, these areas may become prime locations for deer feeding. This, in turn, could contribute to a rise in deer populations and further exacerbate the degradation of forests.

Keywords Dead wood · Effect of harvesting · Bark beetles · Deer

The Stand-based Silvicultural Management System (SSMS, refer to Chap. 7) is a kind of low-impact forestry system that emulates natural disturbance to maintain the structure and functions of mature natural forests as much as possible. Even with the SSMS, however, forestry practices can affect ecological processes to some extent. In this chapter, we focused on the effects of forestry practices on ecological processes.

S. N. Suzuki (✉)
The University of Tokyo Hokkaido Forest, Graduate School of Agricultural and Life Sciences, The University of Tokyo, Furano, Japan

Nakagawa Experimental Forest, Field Science Center for Northern Biosphe, Hokkaido University, Otoineppu, Japan
e-mail: snsuzuki@fsc.hokudai.ac.jp

T. Owari et al. (eds.), *Integrative Forest Management and Silviculture*, Ecological Research Monographs, https://doi.org/10.1007/978-981-95-1017-7_12

12.1 Effects of Forest Management on Dead Woods

It has been reported that the volume of dead wood is lower in managed stands than in unmanaged stands (Siitonen et al. 2000). This is because large trees that have the potential to die in the future are being harvested before they actually die in managed stands. Especially in the SSMS, trees that are likely to die before the next harvesting practice are given priority for logging. As a result, it is expected that the volume of dead wood will naturally decrease in the SSMS. Nakagawa et al. (2001) reported that the volume and cover (projected area) of dead wood on forest floors were almost three times greater in preserved stands than in selection cutting stands (Fig. 12.1). Koike et al. (2009) also reported that the cover of downed logs was much greater in preserved stands than in selection cutting stands (Table 12.1). According to data from the permanent forest plots, the volumetric production of dead wood in the selection cutting stands (2.58 m^2 ha^{-1} $yr.^{-1}$) was almost two-thirds of that in the forest reserves (3.61 m^2 ha^{-1} $yr.^{-1}$) (Fig. 12.2).

Although tree harvesting can reduce the volume of dead wood in selection forests, the volume of dead wood is usually greater in natural forests than in plantation forests (Stokland et al. 2012a, b). The volume of dead wood in well-managed plantations can be less than 10% of that in the same type of natural forest (Siitonen 2001). In a well-managed plantation forest, few trees die, because they are thinned or logged before dying. In particular, large amounts of dead wood are rarely produced, because trees are logged before becoming overmature, which further reduces

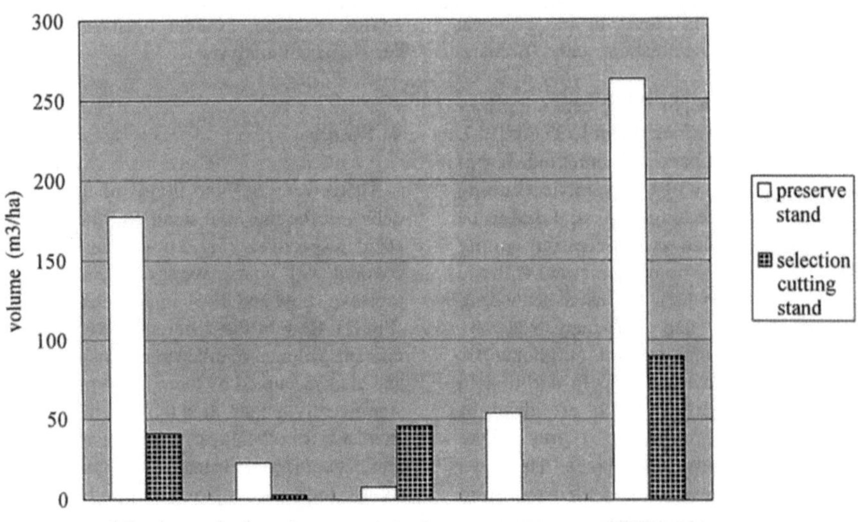

Fig. 12.1 Average volumes of coarse woody debris in the Maeyama Forest Reserve and surrounding selection cutting stands. (Reproduced from Nakagawa et al. 2001, with the publisher's permission)

Table 12.1 Cover of coarse woody debris (m^2 ha^{-1}) in a selection cutting stand and a preserved stand in the UTHF (Translated from Koike et al. 2009)

Types of CWD	Selection cutting stand (Plot No. 5137)	Preserved stand (Plot No. 5146)
Standing dead[a]	1.10	2.92
Dowed logs	3.25	20.42
Stumps	16.96	20.17

[a]The cover of standing dead trees was calculated as the sum of the basal area

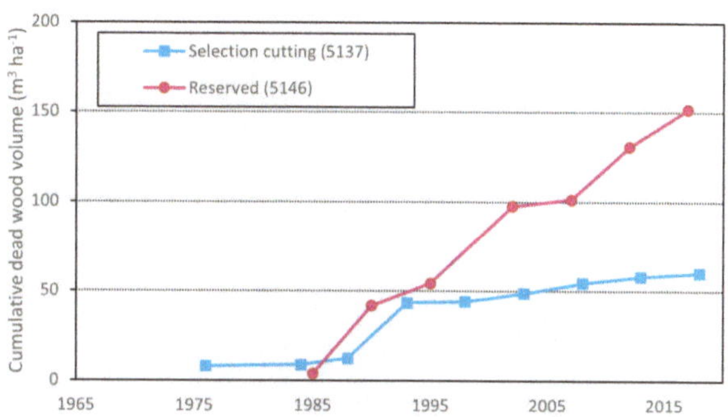

Fig. 12.2 Cumulative volumetric production of dead wood in a selection cutting stand (Plot No. 5137) and a reserved stand (Plot No. 5146). (Data source: UTHF)

the volume of dead wood on the forest floor, because larger trees decay more slowly than smaller trees. Therefore, compared with plantation forests, selection systems can produce large amounts of dead wood of a variety of sizes. In the case of the UTHF, the dead wood production of plantation and managed secondary forests was significantly lower than that of natural selection forests and unmanaged forests (Fig. 12.3a). In particular, the production of large dead wood (DBH > 30 cm) was much lower in the plantation and managed secondary forests than in the other forests (Fig. 12.3b).

The abundance of dead wood can affect a variety of ecosystem components. First, downed logs serve as nursery logs (Harmon and Franklin 1989). Fallen logs provide safe sites for tree seedlings due to reduced competition with other understory species (Harmon and Franklin 1989) and a lower density of pathogenic fungi (O'Hanlon-Manners and Kotanen 2004) compared with soil. In particular, the regeneration of spruce species (genus *Picea*) is largely dependent on nursery logs (Kubota et al. 1994; Kubota and Hara 1996; Takahashi 1997), because spruce seedlings are highly susceptible to soil fungal diseases, such as "snow blight" (see Chap. 14). In addition, the tops of nurse logs serve as refuges from the suppression of *Sasa* dwarf bamboo, because the rhizome of *Sasa* dwarf bamboo rarely climes up downed logs. Therefore, a decrease in dead wood can affect the regeneration of trees. Nakagawa et al. (2001) reported that the distribution of spruce seedlings was strictly limited to dead wood surfaces, and the density of the seedlings was greater in preserved stands than in selection cutting stands in the UTHF (Fig. 12.4). Okada et al. (2019) demonstrated that downed logs also contributed to the biodiversity of herbaceous species on forest floors.

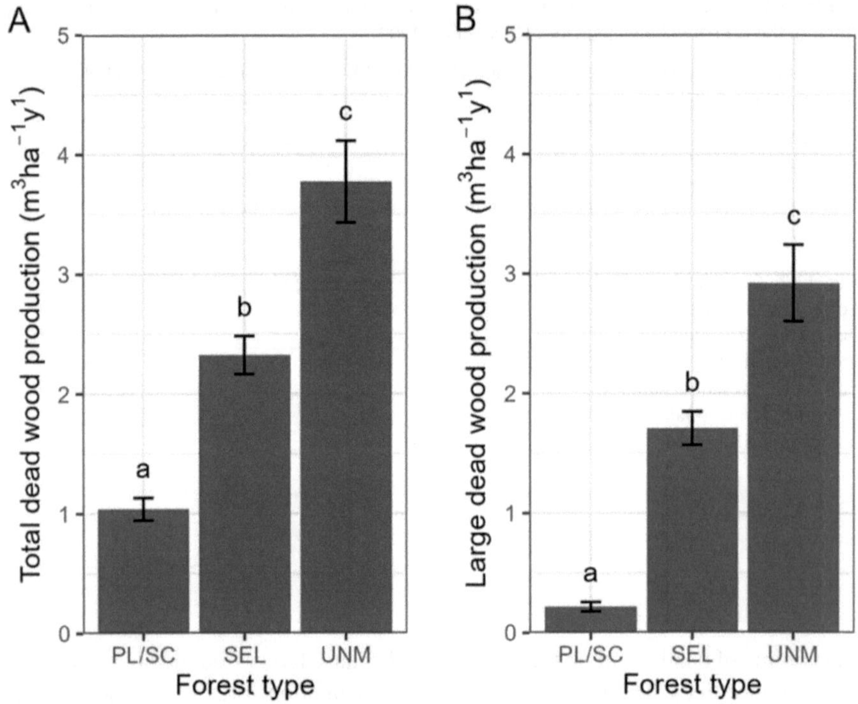

Fig. 12.3 Annual dead wood production rate in plantation and managed secondary forests (PL/SC), selection forests (SEL), and unmanaged forests (UNM) in the UTHF. (**a**) Total dead wood production. (**b**) Dead wood production for large trees (DBH > 30 cm). Managed secondary forests were regenerated after fires in the 1910s and were harvested for thinning. The mean ± SE values are shown. Different lowercase letters above the error bars indicate statistically significant differences at the 5% level. (Data source: UTHF)

Dead woods are also important for the biodiversity and food webs of forest organisms. Dead woods are fed by saproxylic insects such as bark beetles and longhorn beetles and are decayed by fungal species. These wood-decaying fungi are fed upon by other fungivorous insects. These insects, in turn, are fed by higher trophic organisms, that is, predator invertebrates and vertebrates. This kind of food web from deadwoods is called a "saproxylic food web" (Stokland et al. 2012a, b).

The amount of dead wood can affect the abundance of dead-wood-dependent species. For instance, many woodpecker species are associated with dead trees that are used for foraging, roosting, or nesting, and positive correlations between the amount of dead wood and the abundance of woodpecker species have been reported in Europe (Czeszczewik and Walankiewicz 2006; Czeszczewik et al. 2013, 2015; Kosiński et al. 2018). The black woodpecker (*Dryocopus martius*), an endangered species (categorized "Vulnerable (VU)" in Japan), demonstrates a notably high population density within the UTHF (Arisawa 1991). Although this heightened prevalence can be partly attributed to the abundant presence of large-sized Todo firs

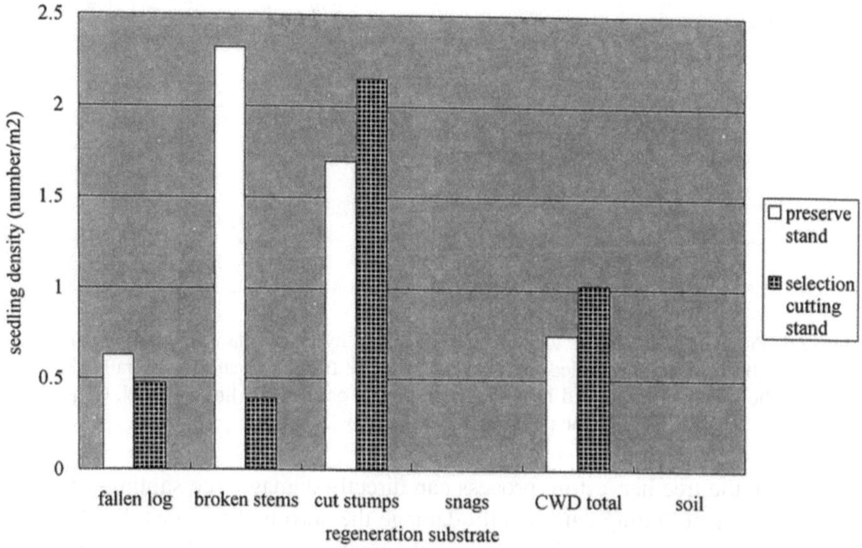

Fig. 12.4 Average seedling density and regeneration substrata of *Picea jezoensis* in the Maeyama Forest Reserve (preserve stand) and surrounding selection cutting stand. The seedling density was calculated as the sum of the seedling density at specific substrata and the total cover of the substrata per hectare of forest floor. (Reproduced from Nakagawa et al. 2001, with the publisher's permission)

in the UTHF, which serve as favorable nesting trees for the species (Arisawa 1991), the substantial presence of dead wood in the selection forest also potentially contributes to creating a suitable habitat for this endangered species. In the context of habitat preservation, a comprehensive review of thresholds for dead wood suggests that maintaining a range of 30–50 m^3 per hectare is critical for fostering biodiversity within boreal montane mixed forests (Müller and Bütler 2010). Notably, the quantity of dead wood in the selection stands exceeded this proposed threshold within the UTHF (Fig. 12.1), suggesting that the UTHF plays a crucial role in fostering the delicate equilibrium between wood production and biodiversity conservation.

12.2 Effects of Tree Harvesting on Forest Floor Conditions

Tree harvesting can have a large impact on forest floor conditions. First, canopy opening resulting from tree harvesting leads to improvements in light environments. An improved light environment can provide opportunities for regeneration of tree species, while it can also cause propagation of Sasa dwarf bamboos. The propagation of Sasa dwarf bamboos, in turn, limits the regeneration of trees. Whether tree harvesting positively or negatively affects the regeneration of trees largely depends on the density of the dwarf bamboos and advanced regeneration prior to tree harvesting.

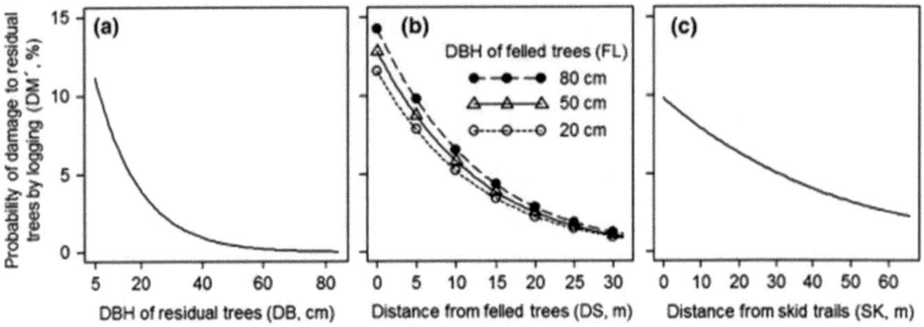

Fig. 12.5 Probability of damage to residual trees caused by logging as a function of (**a**) residual tree DBH, (**b**) distance from felled trees, and (**c**) distance from the nearest skid trail in selection stands in the UTHF, as predicted by an individual-level generalized linear model. (Reproduced from Tatsumi et al. 2014, with the publisher's permission)

Second, the tree harvesting process can directly damage tree saplings (Surakka et al. 2011). Tree felling can directly damage the surrounding trees. Furthermore, the skidding of felled trees also damages the trees near skid trails, especially small ones (i.e., saplings). Tatsumi et al. (2014) reported that 3% of residual trees (larger than 5 cm in DBH) after harvesting died due to damage from the harvesting process in the UTHF and that smaller trees and trees closer to felled trees and skidding trails were more likely to die (Fig. 12.5). Although they did not directly evaluate the effects of harvesting on the survival of saplings (DBH < 5 cm), their results imply that smaller trees have a greater probability of mortality during the process of harvesting.

12.3 Effects of Selection Cutting on Forest Composition

Changes in structure and composition of living trees in selection cutting stands in the UTHF have also been documented (Yamamoto et al. 1997; Kitabatake et al. 2003; Tatsumi et al. 2010). Notably, a significant decrease in stem density of small-sized conifers has been observed (Yamamoto et al. 1997, Kitabatake et al. 2003, Tatsumi et al. 2010, Fig. 12.6). This decline may be partly related to changes in dead wood quantity and understory condition mentioned earlier. Additionally, selection cutting has often targeted conifers (Yamamoto et al. 1997). Consequently, in the first Natural Forest Working Group (refer to Chap. 2) in the UTHF, the stem density of *Abies sachalinensis* decreased from 378 stems ha^{-1} in 1964–1968 to 312 stems ha^{-1} in 1996–2005, and *Picea jezoensis* also decreased from 63 to 47 stems ha^{-1} during the same period, while the stem density of broadleaves increased from 410 to 432 stems ha^{-1} (Tatsumi et al. 2010). Relative basal area (UTHF, unpublished data) or relative stock volume (Tatsumi et al. 2010) of conifers has also tended to decline over the past half-century.

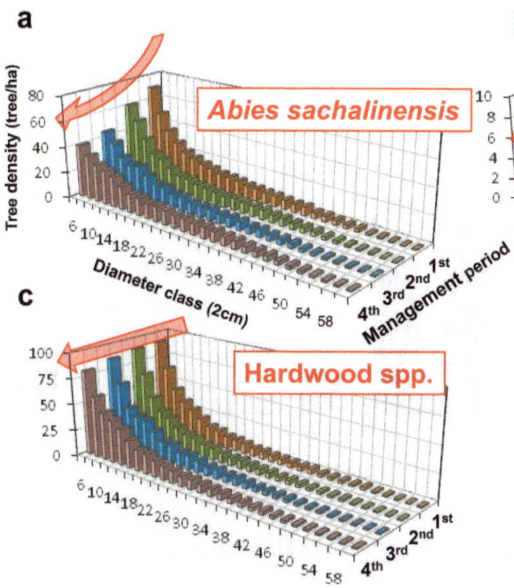

Fig. 12.6 The DBH distribution of Conifers (**a.** *Abies sachalinensis*, **b.** *Picea jezoensis*) and broadleaves (**c**) in the first Natural Forest Working Group in the UTHF. The first, second, third, fourth management period is corresponding to 1964–1967, 1968–1980, 1986–1995, and 1996–2005, respectively. The DBH distribution was estimated by measurements taken in inventory plots across an entire area of the first NFWG. (Redrawn after Tatsumi et al. 2010 by the original authors with the publisher's permission)

The decline in conifers may negatively impact their own regeneration. Sasa dwarf bamboo is typically suppressed under the canopy of evergreen conifers, particularly medium-sized trees (33–45 cm in DBH), leading to lower density and height of dwarf bamboo beneath these canopies (Tatsumi and Owari 2013). Thus, the reduction in evergreen conifers may facilitate the spread of dwarf bamboo, which can diminish opportunities for the regeneration of tree species, especially conifers. To counteract this negative cycle, the UTHF is now focusing on increasing the proportion of broadleaves as a target in selection cutting.

12.4 Effects of Tree Harvesting on Insects

Tree harvesting affects the dynamics of saproxylic invertebrates. In particular, bark and ambrosia beetles often increase in harvested stands and kill surrounding trees (Koizumi 1977). In the UTHF, an increase in the prevalence of the eight-toothed spruce bark beetle *Ips typographus japonicus* and its infection of spruce trees after selection cutting have been reported (Nakayama et al. 1991, Iguchi et al. 1993, Iguchi 2023, refer also to Chap. 14). The mortality rates of spruce trees increased for several years after harvesting (Iguchi et al. 1993, Fig. 12.7). Trees near logging

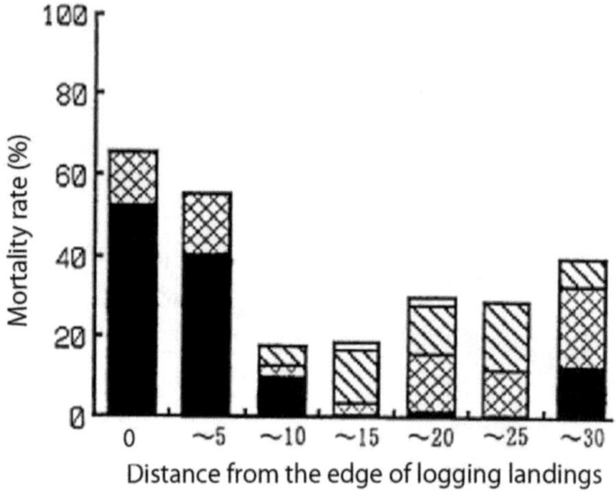

Fig. 12.7 Relationships between the mortality rate of spruce trees and the distance (meters) from the edge of the logging landing 2–5 years after selection cutting in the UTHF. Black area, 2 years; cross shading, 3 years; slash shading, 4 years; white area, 5 years after logging. (Modified after Iguchi et al. 1993 by Satoshi N. Suzuki with the publisher's permission)

landings are more likely to be attacked by the spruce bark beetles (Nakayama et al. 1991; Iguchi et al. 1993).

12.5 Effects of Tree Harvesting on Vertebrates

Deer favor, and are drawn to, open areas around forests, because these sites offer an abundance their preferred diets, such as grass and herbs. Tree harvesting operations create new open sites such as small gaps from felling, log pathways, and logging landings, which also provide leaves of logged trees as food for deer. It is documented that deer population tends to increase in harvested areas (Chang et al. 1995; Smolko et al. 2018). While the density of sika deer (*Cervus nippon*) is not high in the UTHF (refer to Chap. 14), they are attracted to recent clear-cut or planted sites and browse not only grass and herbs but also the leaves and bark of regenerating or planted trees.

In a recently clear-cut area, a significant proportion of the sprouted shoots of *Tilia japonica* and *Tilia maximowicziana* were browsed by deer up to 13 years after clearcutting in the UTHF (Tokuni et al. 2008, 2020). This observation implies that deer are drawn to clear-cut sites and actively engage in extensive browsing of plants within those areas. Additionally, deer density appears to be significantly high in or around sparse stands that are densely covered by *Sasa* dwarf bamboo, resulting in an elevated risk of bark stripping on trees (Suzuki et al. 2011). If tree harvesting leads to the creation of more newly clear-cut sites or sparse forests, these areas may become prime locations for deer feeding. This, in turn, could contribute to a rise in deer populations and further exacerbate the degradation of forests.

The effects of selection cutting on vertebrates in UTHF are unclear. This is because the impacts of selection cutting on forest structure are significantly smaller than those of clearcutting. Large mammals may not care about such slight differences in forest structure. Although differences in the abundance of large trees or dead wood may affect the abundance of small mammals and birds that depends on these features (as mentioned in12.1), such effects have rarely been assessed. However, there have been observations of differences in bat activity among unmanaged forests, forests managed with selection cutting, secondary forests, and plantation forests have been observed in the UTHF (Fukui, unpublished data). Selection cutting generally had a negative effect on bat activity, though it positively affected the activity of certain species. Although the reasons for this are not fully understood, their results suggest that selection cutting might affect bat activity differently across different species.

12.6 Remarks

In conclusion, large-scale practices such as clearcutting and plantations can significantly impact ecosystem processes, notably by reducing the amount of dead wood, which is crucial for tree regeneration and the organisms involved in saproxylic food webs. While selection cutting under the SSMS can also decrease dead wood compared to unmanaged forests, the dead wood in the UTHF is considered sufficient to sustain the biodiversity of dead-wood-dependent species. However, the regeneration of *Picea* species, wood-dependent conifers, is significantly reduced in selection cutting stands due to decreased levels of dead wood quantity. Additionally, even under the selection system, tree harvesting can have a profound impact on forest floor conditions, especially by damaging tree saplings, leading to a decline in the stem density of conifers. Tree harvesting practices can also influence the dynamics of bark beetles. If tree harvesting results in the creation of more clear-cut sites or sparse forests, these areas may attract deer, leading to an increase in deer populations and further forest degradation. Therefore, careful consideration of these impacts is essential in developing sustainable forest management practices that balance tree harvesting with the preservation of biodiversity and ecosystem health.

References

Arisawa H (1991) Breeding nest density and the characteristics of nest tree of black woodpecker, *Dryocopus martius martius*. Bullet Univ Tokyo Forests 84:21–37

Chang KT, Verbyla DL, Yeo JJ (1995) Spatial analysis of habitat selection by Sitka black-tailed deer in Southeast Alaska, USA. Environ Manag 19:579–589. https://doi.org/10.1007/BF02471969

Czeszczewik D, Walankiewicz W (2006) Logging affects the white-backed woodpecker *Dendrocopos leucotos* distribution in the Białowieza Forest. Ann Zool Fenn 43:221–227. https://www.researchgate.net/publication/279560356

Czeszczewik D, Walankiewicz W, Mitrus C, Tumiel T, Stański T, Sahel M, Bednarczyk G (2013) Importance of dead wood resources for woodpeckers in coniferous stands of the Białowieża Forest. Bird Conserv Int 23:414–425. https://doi.org/10.1017/S0959270912000354

Czeszczewik D, Zub K, Stanski T, Sahel M, Kapusta A, Walankiewicz W (2015) Effects of forest management on bird assemblages in the Bialowieza forest, Poland. IForest 8:377–385. https://doi.org/10.3832/ifor1212-007

Harmon ME, Franklin JF (1989) Tree seedlings on logs in *Picea-Tsuga* forests of Oregon and Washington. Ecology 70:48–59

Iguchi K (2023) Effects of increased food resources due to a wind-disturbance on reproduction of a spruce bark beetle (*Ips typographus japonicus* NIIJIMA). Bullet Univ Tokyo Forests 149:1–25. http://hdl.handle.net/2261/0002007918. (in Japanese)

Iguchi K, Yamamoto H, Furuta K (1993) The spread of attack of Yezo-spruce by spruce beetles in natural forests after selection cutting. Bullet Univ Tokyo Forests 90:1–5. (in Japanese)

Kitabatake T, Goto S, Takahashi Y, Kasahara H, Inukai M (2003) Effect of selection cutting on population dynamics of *Abies sachalinensis* in a cool temperate mixed forest in Hokkaido, Japan. Japanese J Forest Soc 85:252–258

Koike Y, Inukai H, Fukushi K, Murakawa I, Takahashi K, Inukai S, Owari T (2009) A comparison of structure between a selection cutting stand and a reserved stand in the Tokyo university Forest in Hokkaido. Boreal Forest Res 57:105–107. (in Japanese)

Koizumi C (1977) Cutting operations in the spruce-fir forest in Hokkaido. Bullet Govern Forest Exp Station 297:1–34. (in Japanese)

Kosiński Z, Pluta M, Ulanowska A, Walczak Ł, Winiecki A, Zarębski M (2018) Do increases in the availability of standing dead trees affect the abundance, nest-site use, and niche partitioning of great spotted and middle spotted woodpeckers in riverine forests? Biodivers Conserv 27:123–145. https://doi.org/10.1007/s10531-017-1425-6

Kubota Y, Hara T (1996) Recruitment processes and species coexistence in a subboreal forest in northern Japan. Ann Bot 78:741–748. https://doi.org/10.1006/anbo.1996.0184

Kubota Y, Konno Y, Hiura T (1994) Stand structure and growth patterns of understorey trees in a coniferous forest, Taisetsuzan National Park, northern Japan. Ecol Res 9:333–341. https://doi.org/10.1007/BF02348420

Müller J, Bütler R (2010) A review of habitat thresholds for dead wood: a baseline for management recommendations in European forests. Eurasian J Forest Res 129:981–992. https://doi.org/10.1007/s10342-010-0400-5

Nakagawa M, Kurahashi A, Kaji M, Hogetsu T (2001) The effects of selection cutting of a *Picea jezoensis* and *Abies sacharinensis* in the subboreal forests of Hokkaido, northern Japan. For Ecol Manag 146:15–23

Nakayama M, Furuta K, Takahashi I, Sato Y, Ituti K (1991) Death of Ezo-spruce around log depots after partial cuttings and the dynamics of adult spruce beetles. Bullet Tokyo Univ Forests 84:39–52. (in Japanese)

O'Hanlon-Manners DL, Kotanen PM (2004) Logs as refuges from fungal pathogens for seeds of eastern hemlock (*Tsuga canadensis*). Ecology 85:284–289

Okada M, Hirao T, Kaji M, Goto S (2019) Role of fallen logs in maintaining the species diversity of understory vascular plants in a mixed coniferous and broad-leaved forest in Hokkaido, northern Japan. For Ecol Manag 448:249–255

Siitonen J (2001) Forest management, coarse woody debris and saproxylic organisms: Fennoscandian boreal forests as an example. Ecol Bull 49:11–41. http://www.jstor.org/stable/20113262

Siitonen J, Martikainen P, Punttila P, Rauh J (2000) Coarse woody debris and stand characteristics in mature managed and old-growth boreal Mesic forests in southern Finland. For Ecol Manag 128:211–225. https://doi.org/10.1016/S0378-1127(99)00148-6

Smolko P, Veselovská A, Kropil R (2018) Seasonal dynamics of forage for red deer in temperate forests: importance of the habitat properties, stand development stage and overstorey dynamics. Wildl Biol 2018:1–10 wlb.00366. https://doi.org/10.2981/wlb.00366

Stokland J, Siitonen J, Jonsson B (2012a) Biodiversity in dead wood (ecology, biodiversity and conservation). Cambridge University Press, Cambridge. https://doi.org/10.1017/CBO9781139025843

Stokland JN, Siitonen J, Jonsson BG (2012b) Biodiversity in dead wood. Cambridge University Press, Cambridge

Surakka H, Sire'n M, Heikkinen J, Valkonen S (2011) Damage to saplings in mechanized selection cutting in uneven-aged Norway spruce stands. Scand J For Res 26:232–244

Suzuki M, Fujiwara A, Kamoda S, Maehara T, Saito H, Matsui M, Iguchi K, Kaji M, Kamata N (2011) Factors affecting the probability of bark-stripping by deer in a natural forest: an analysis based on tree characteristics, deer habitat utilization and environmental conditions. Japanese J Forest Soc 93:213–219. (in Japanese with English summary)

Takahashi K (1997) Regeneration and coexistence of two subalpine conifer species in relation to dwarf bamboo in the understorey. J Veg Sci 8:529–536. https://doi.org/10.2307/3237203

Tatsumi S, Owari T (2013) Modeling the effects of individual-tree size, distance, and species on understory vegetation based on neighborhood analysis. Can J For Res 43:1006–1013

Tatsumi S, Owari T, Yamamoto H, Shiraishi N (2010) Change of the stand structure by natural selection forestry in 42 years—a case study in the Tokyo university Forest in Hokkaido. Kanto J Forest Res 61:54–56. (in Japanese)

Tatsumi S, Owari T, Kasahara H, Nakagawa Y (2014) Individual-level analysis of damage to residual trees after single-tree selection harvesting in northern Japanese mixedwood stands. J For Res 19:369–378

Tokuni M, Kasahara H, Kimura N, Shimizume M, Nakagawa Y, Goto S (2008) The occurrence of sprouts before and after harvesting in *Tilia japonica* and *T. Maximowicziana*. Boreal Forest Res 56:59–61. (in Japanese)

Tokuni M, Fukui D, Oikawa N, Ogawa H (2020) The occurrence of sprouts in *Tilia japonica* and *T. maximowicziana*: a comparison between 1 and 13 years after harvesting. Boreal Forest Res 68:57–59. (in Japanese)

Yamamoto H, Ohashi K, Michigami A, Shibano S, Iwamoto S, Inukai H, Sato I (1997) Stand structure of mixed-species stands (III) the change of stand structure with selection cutting operations. Trans Japanese Forestry Soc 108:91–94. (in Japanese)

Chapter 13
Forest Recovery Process and Restoration After Large-Scale Disturbances

Satoshi N. Suzuki and Toshiaki Owari

Abstract In Hokkaido, windstorms including typhoons are major disturbances, notably Typhoon Marie in 1954 and Typhoon Thad in 1981, which caused extensive damage to the University of Tokyo Hokkaido Forest (UTHF). Conifers are more susceptible to wind damage than broadleaves. Post disturbance, the UTHF employed salvage logging and reforestation. Conifer planted stands generally successfully grew up although invaded broadleaves often outcompete planted conifers. Naturally regenerated stands after soil scarification were characterized as *Betula* species with too high of a density. Naturally regenerated stands without any postdisturbance practices were dominated by mixed conifer-broadleaf forests. Salvage logging also impacts ecosystems by reducing the habitat available for saproxylic organisms and altering carbon stocks. Large-scale human-caused wildfires in the 1910s created a vast area of secondary forests, which are currently dominated by *Betula maximowicziana.* Some of the Norway spruce plantations established after these fires have also matured into old-growth stands. Snow damage, particularly in young forests, further stress the forest, with wet snow bending or breaking tree stems. Due to climate change, the frequency and severity of these disturbances are expected to increase, prompting the UTHF focus on developing forests that are more robust and resilient to such disturbances.

Keywords Salvage logging · Snow damage · Typhoon · Wildfire

S. N. Suzuki (✉)
The University of Tokyo Hokkaido Forest, Graduate School of Agricultural and Life Sciences, The University of Tokyo, Furano, Japan

Nakagawa Experimental Forest, Field Science Center for Northern Biosphere, Hokkaido University, Otoineppu, Japan
e-mail: snsuzuki@fsc.hokudai.ac.jp

T. Owari
The University of Tokyo Hokkaido Forest, Graduate School of Agricultural and Life Sciences, The University of Tokyo, Furano, Japan
e-mail: owari@g.ecc.u-tokyo.ac.jp

T. Owari et al. (eds.), *Integrative Forest Management and Silviculture*,
Ecological Research Monographs, https://doi.org/10.1007/978-981-95-1017-7_13

13.1 Windstorm: Impacts of Typhoons

13.1.1 Damages by Typhoons

The primary agent of large-scale disturbance in Hokkaido is strong winds from typhoons and other windstorms. Every few decades, typhoons cause significant forest blowdowns in Hokkaido. In the University of Tokyo Hokkaido Forest (UTHF), several instances of large-scale wind-induced disturbances have been recorded.

The most significant wind disturbance in Hokkaido occurred during Typhoon Marie in 1954 (T5415, named "Touyamaru typhoon" in Japan), which destroyed 26,900,000 m^3 of trees over an area of 760,000 ha in Hokkaido (Watanabe et al. 1990) and 137,690 m^3 of trees in the UTHF (Hirata and Maezawa 1956). The most significant wind disturbance in the UTHF was caused by Typhoon Thad in 1981 (T8115, the Typhoon No. 15 in 1981). Although the damage caused by Typhoon Thad in 1981 in Hokkaido (3,709,000 m^3 of trees in an area of 58,780 ha) was less than that caused by Typhoon Marie in 1954, it destroyed 755,430 m^3 or more of trees in the area of 10,649 ha in the UTHF (Watanabe et al. 1990, refer to Fig. 2.9), surpassing the damage caused by Typhoon Marie in that area. The damaged area was equated to 46.6% of the UTHF forest area (Fig. 13.1). Additionally, 43,863 m^3 of trees were damaged by insect infestations following Typhoon Thad. Typhoon Helen in 1972 (T7220) also damaged 25,000 m^3 of trees (Tokyo University Forest in Hokkaido 1973).

Topographic factors are important for vulnerability to wind. The extent of wind exposure in a forest depends largely on the slope's orientation to the wind direction and the forest's position on a slope (ridge or valley). Forests on ridges are generally more vulnerable to wind. Up-valley winds (from the valley-mouth direction) can also damage trees in valleys. Slope steepness is another crucial factor. In the case of Typhoon Thad in the UTHF, forests on south-facing slopes, gently sloping terrain aligned with the wind direction, showed the highest probability of windthrow (Miura et al. 1984; Shibano et al. 1990).

Conifers are generally more susceptible to strong winds (Gardiner 2021). During Typhoon Thad, 25.3% of the total conifer volume was damaged, compared to only 10.6% of broadleaves in the UTHF (Watanabe et al. 1990). During Typhoon Marie, these percentages were 4.8% and 1.5%, respectively (Hirata and Maezawa 1956).

13.1.2 Post-windthrow Management

Following the two severe wind disturbances, most of the wind-damaged trees were removed from the forest through so-called salvaged logging. Salvage logging aimed not only at recovering from economic loss by selling the salvaged trees but also at preventing insect outbreaks, such as spruce bark beetles, from the downed trees. After Typhoon Thad in 1981, 755,430 m^3 of stems were removed from the UTHF,

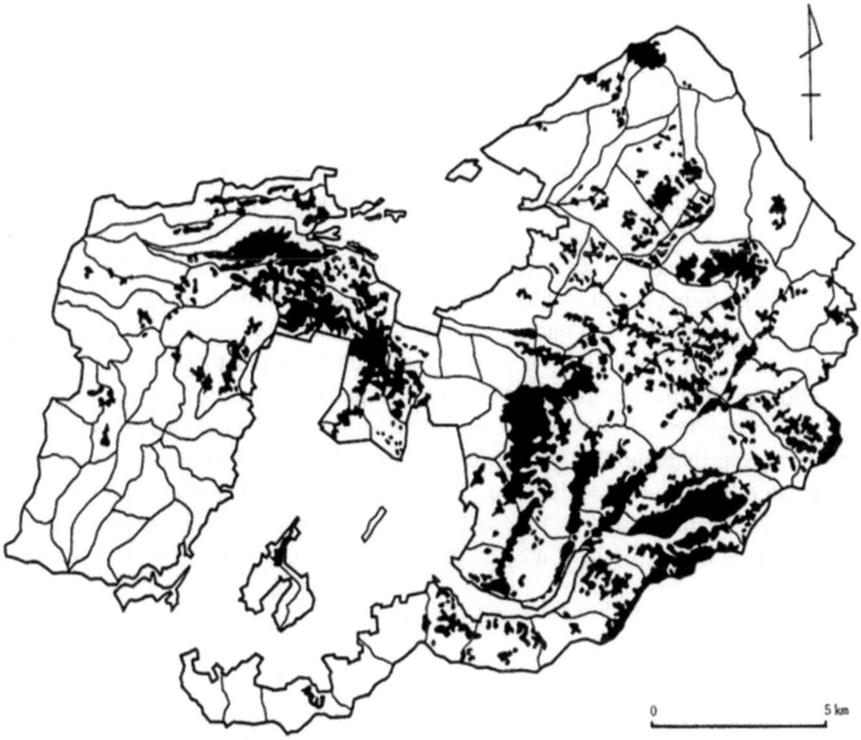

Fig. 13.1 Area severely damaged (shown in black) by Typhoon Thad in UTHF in 1981. (Reproduced from Watanabe et al. 1990, with the publisher's permission)

equivalent to 18% of the growing stock and twelve times of the average annual harvesting volume of the UTHF at that time (Watanabe et al. 1990).

Postdisturbance restoration efforts were actively implemented following the salvage logging after Typhoon Thad although areas with abundant natural regeneration were left untreated after the salvage logging. Between 1981 and 1986, 483 hectares were planted, primarily with Sakhalin fir *Abies sachalinensis* and Sakhalin spruce *Picea glehnii*. Additionally, 307 hectares were just scarified to remove *Sasa* dwarf bamboo and facilitate natural regeneration. In additional areas (86 hectares), seeding was also conducted after scarification to compensate for the shortage of natural seed dispersal, including seeding of Japanese oak *Quercus crispula,* (62 ha) and mixture of monarch birch *Betula maximowicziana* and Yezo spruce *Picea jezoensis* (24 hectare). Study plots were established in these reforested areas to monitor the forest recovery process.

After Typhoon Thad, the planting density was generally reduced to 1500 seedlings/ha, compared to the traditional planting density of 2500–3000 seedlings/ha (Watanabe et al. 1990), to encourage the natural regeneration of other species, particularly broadleaves, for the restoration of natural mixed conifer-broadleaf forests. Before planting, scarification process as site preparation was carried out in a striped pattern using either a bulldozer or a rake dozer, with 3–4-m-wide strips being scarified and similar-sized strips left unscarified (residual strips). Seedlings were then planted in the scarified strips. At most of the planted sites, many broadleaves regenerated in the residual strips (Okamura et al. 1995; Asama et al. 1996; Shimizume et al. 2002; Okamura et al. 2006; Matsui et al. 2008). However, naturally regenerated species such as *Betula maximowicziana* often outcompete and suppress the planted conifers (Shimizume et al. 2002, Okamura et al. 2006, Matsui et al. 2008). Shimizume et al. (2002) reported that 45% of planted conifers exhibited symptoms of some kind of damage due to suppression by broadleaves at a planted site 15 years after planting. Okamura et al. (2006) found that early thinning of broadleaves significantly improved the growth and quality of both planted conifers and high-value broadleaves like *B. maximowicaiana* at planted sites 22 years after planting. Approximately 30 years after planting, the planted conifers began to mix with naturally regenerated broadleaves, forming conifer-broadleaf mixed forests (Suzuki 2020). In particular, *A. sachalinensis* has grown well and become codominant with broad leaves in some stands, forming mixed conifer-broadleaf forests (Fig. 13.2).

Fig. 13.2 Sakhalin fir plantation after the windthrow in 1981. (Photo by Satoshi N. Suzuki, July 16, 2024)

As noted in Chap. 10, the scarified areas that were left unplanted, including those with only scarification and those with seeding following scarification, have predominantly transitioned to *Betula*-dominant stands. In areas that received only scarification, which were mostly located at higher elevations, *B. ermanii* became particularly dominant. The stem density of *B. ermanii* in these areas often exceeded 10,000 stems per hectare, sometimes reaching over 200,000 stems per hectare a decade after scarification (Okamura et al. 1995). However, due to the very high density, the stem growth of the *Betrula* trees was very limited, resulting in slender stands unsuitable for timber production (Fig. 13.3a). Most seeding sites were similarly dominated by slender stems of *Betula*, including those regenerated from seeded seeds, although some seeded Japanese oak *Q. crispula* still survive under the birch canopy. Consequently, scarification and seeding treatments were seen as unsuccessful for restoring mixed conifer-broadleaf forests without any further management. Previous studies in Hokkaido have suggested that management operations, such as bush clearing or thinning, in the early stage of stand development improved the growth of *B. ermanii* in scarified areas (Kikuzawa 1989; Miyoshi et al. 2014; Sano and Shibuya 2015). To address this issue, the UTHF initiated a trial of stem density control in 2023 by implementing strip-clearcutting in a grid pattern, cutting in two directions to improve the growth of *B. ermanii* and facilitate the regeneration of other species (Fig. 13.3b).

Mixed conifer-broadleaf stands developed in some areas where only salvage logging was conducted (Kisanuki et al. 1996, Suzuki 2020, Li et al. 2023; see also Fig. 5.6). This occurred because the advanced regeneration of conifers, which already existed on the forest floor before the disturbance, contributed to the recovery of the stands. The sites disturbed by Typhoon Marie in 1954 and where only salvage logging was done are also now semimature mixed conifer-broadleaf stands. Sprouting from downed logs or cut stamps also occurred for some broadleaf species, such as *Tilia japonica*, *Sorbus alnifolia*, and *Acer* spp. (Kisanuki et al. 1996), which largely contributed to the recovery of forest stands after the typhoon. These findings indicate that disturbance legacies such as advanced saplings, downed logs and stumps are important for the restoration of mixed conifer-broadleaf forests. In other words, scarification, which removes such disturbance legacies, negatively impacts the restoration of mixed conifer-broadleaf forests.

Salvage logging also impacts forest ecosystems after windstorms. Removing downed trees can reduce habitats and resources for saproxylic organisms, such as insects and fungi (refer to Chap. 12), affecting higher trophic levels like woodpeckers. The removal of downed logs can also alter the carbon stocking function of forests. In cool climates, such as boreal and subalpine regions, as the decomposition rate of dead wood is very slow, dead wood acts as a major carbon stock. Therefore, dead wood can contribute to early carbon stock recovery in disturbed forests by offsetting reduced living carbon stocks (Suzuki et al. 2019; Hotta et al. 2019, 2021,

Fig. 13.3 (**a**) Secondary *Betula ermanii* stand regenerated after salvage logging and scarification treatment following the windthrow in 1981. (**b**) Strip-clear cutting of the *B. ermanii* stand. Stems were uprooted or snapped (rather than cut) using an excavator equipped with a grapple, creating 5-m-wide clear-cut rows in a grid pattern. In these rows, *Sasa* dwarf bamboo was also removed by the machine to facilitate natural regeneration of trees. (Photo by (a) Satoshi N. Suzuki, May 29, 2024, (b) the UTHF, August 31, 2023, licensed under CC BY-NC-ND 4.0)

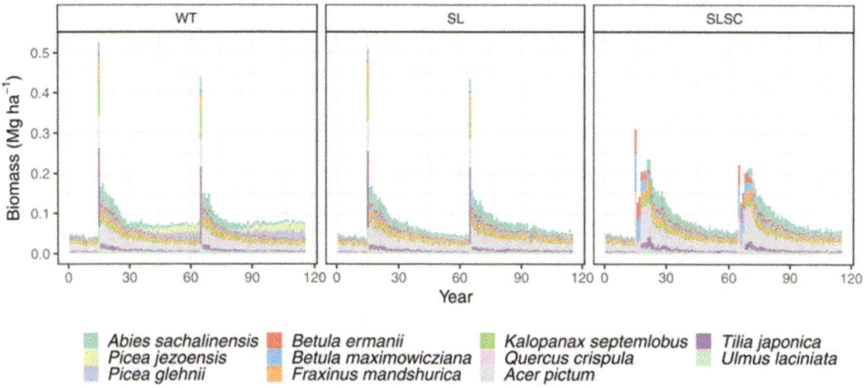

Fig. 13.4 Simulated species composition of cohorts established each year in stands where windthrows occurred in years 15 and 65. WT: scenario in which dead wood generated by windthrow is left intact; SL: scenario in which dead wood generated by windthrow is salvaged; SLSC: scenario in which dead wood generated by windthrow is salvaged, followed by scarification. Note that the biomass of *Picea jezoensis* (filled in yellow) was too low to distinguish in both SL and SLSC. (Reproduced from Hotta et al. 2021, with the publisher's permission)

2023). Additionally, removing downed logs can reduce regeneration opportunities for nursery log-dependent species such as *Picea* species. According to the simulation study of Hotta et al. (2021), salvage logging under a scenario of repeated wind disturbance could significantly decrease the biomass of *Picea* species (Fig. 13.4).

13.2 Fire

Currently, wildfires are relatively rare disturbances in Hokkaido due to the humid climate and fire management practice. However, during the early twentieth century, a period of increased activity by settlers from mainland Japan, wildfires occurred more frequently, mainly due to human activity. In the UTHF, two significant wildfires occurred in 1911 and 1914. The 1911 wildfire originated outside of the UTHF and burned an area of 3680 hectares in the western part of the UTHF (Refer to Chap. 2), and the 1914 wildfire burned approximately 1400 hectares, although there was considerable overlap between the two burned area (Tokyo University Forests 1950, Fig. 13.5). In total, an area of 3900 hectares of forest was burned in the early 1910s. Extensive afforestation efforts followed these wildfires, mainly involving the planting of Norway spruce, with some planting of Japanese larch and Scots pine in the burnt areas. Some of these planted trees still remain and have matured into more than 100-year-old plantations (refer to Chap. 10). However, a large portion of the plantations were unsuccessful, and the originally planted species were gradually

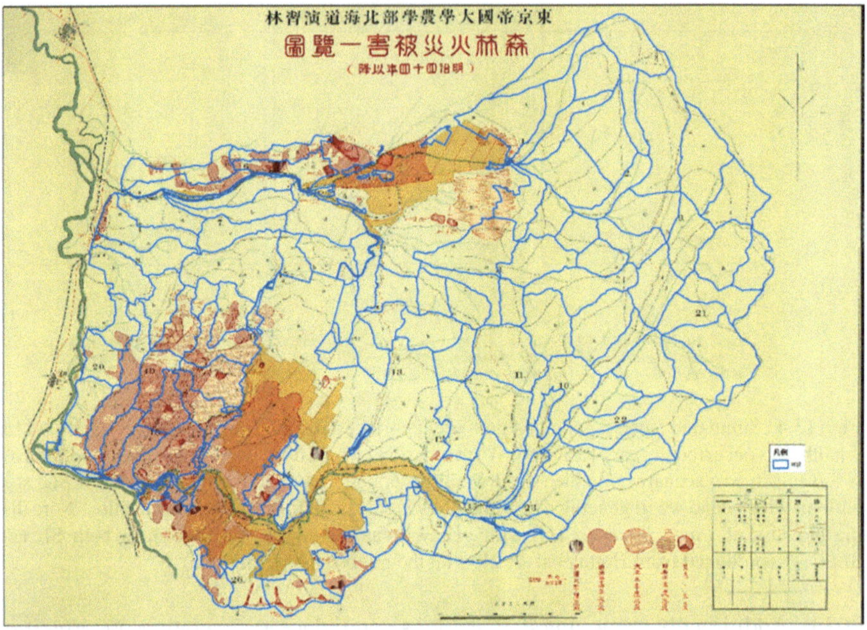

Fig. 13.5 Map of areas burned by wildfires by 1929 in the UTHF. Areas with solid diagonal lines from top right to bottom left were burned in 1911; areas with broken diagonal lines from top left to bottom right were burned in 1914; areas with broken horizontal lines were burned in 1929; areas with other patterns were burned in years other than these. Blue lines indicate the boundaries of current forest compartments. (Source: Digital Archives of the University of Tokyo Hokkado Forest, https://electra.lib.a.u-tokyo.ac.jp/uthf/jpeg/hokuen21.jpg, modified by the UTHF)

replaced by naturally regenerated broadleaf species. No afforestation efforts were conducted at the sites with abundant natural regeneration. Today, most of the burnt sites, including both the unplanted area and the unsuccessful plantation, are dominated by *Betula* species, especially *B. maximowicziana* (Fig. 13.6). *Quercus crispula* also dominated in some stands (Fig. 13.7). Even in the *Betula* dominant stands, the dominance of *Q. crispula* and *A. sachalinensis*, late successional species, gradually increased (Fig. 13.8).

These hardwood trees have become valuable resources for timber production (refer to Chap. 8). Because *B. maximowicziana* is an especially high-value species, density control has been deliberately conducted to facilitate its growth. The target density was set at 29 trees per hectare, based on the crown size of *B. maximowicziana* individuals that have superior traits as timbers in natural mature stands in the UTHF (Takada et al. 1988). Thinning significantly improved the growth of *B. maximowicziana* (Fig. 13.9).

Fig. 13.6 A *Betula maximowicziana* stand regenerated after a fire in 1910s. (Photo by Makoto Ishiguro, Septemer 13, 2023, used with permission from the UTHF, licensed under CC BY-NC-ND 4.0)

Fig. 13.7 A *Quercus crispula* stand regenerated after a fire in 1910s. (Photo by Satoshi N. Suzuki, July 17, 2024)

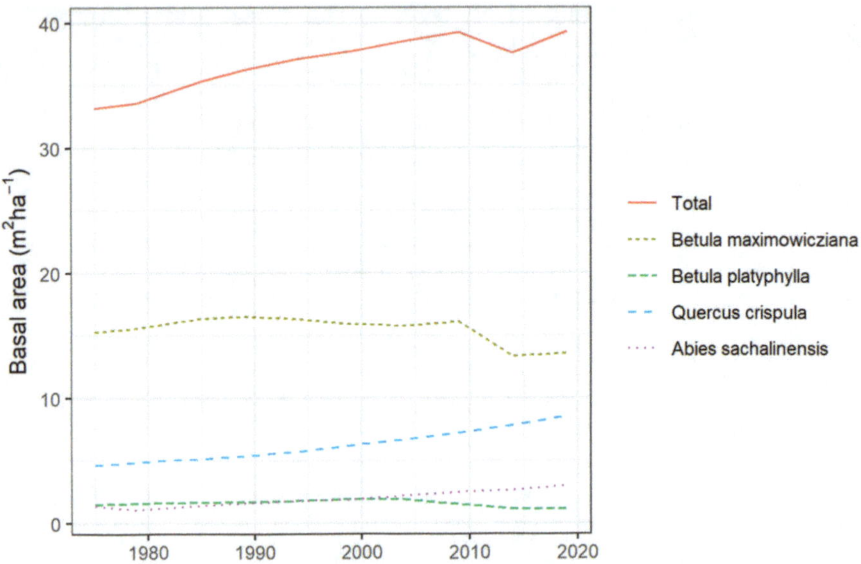

Fig. 13.8 Changes in the stand basal area of *Betula maximowicziana*, *B. platyphylla* var. *japonica*, *Quercus crispula*, and *Abies sachalinensis* in the secondary forest regenerated after the fire in 1911. (Plot No. 5234, Data source: the UTHF)

Uchiyama et al. (2006) reported that some populations of *B. maximowicziana* in fire-disturbed areas showed genetic signs of past population bottlenecks (lower effective population sizes estimated by genetic markers). This may imply that the population of *B. maximowicziana* steeply increased after the fire disturbance.

13.3 Snow Damage

Forest disturbance caused by snow is a relatively small but frequently occurs in high-latitude and high-elevation areas (Nagel et al. 2017; Duperat et al. 2020; Suvanto et al. 2021), especially in plantation forests and young secondary forests, where tree shapes are relatively slender. Snow in winter is typically dry and light in central Hokkaido, which is not problematic for trees. However, when a large amount of snow falls in autumn or spring, wet snow can cause heavy snowpack accumulation on tree crowns, which can cause the bending or breakage of stems (Fig. 13.10a).

In the UTHF, large-scale severe snow damage was reported in 1970 and 1972, affecting 5000 m^3 and 10,000 m^3 of trees, respectively (Tokyo University Forest in Hokkaido 1973). Wet snow on October 16, 2013, also caused significant damage to some forest stands, primarily affecting larch and broadleaf plantations. Wet snow

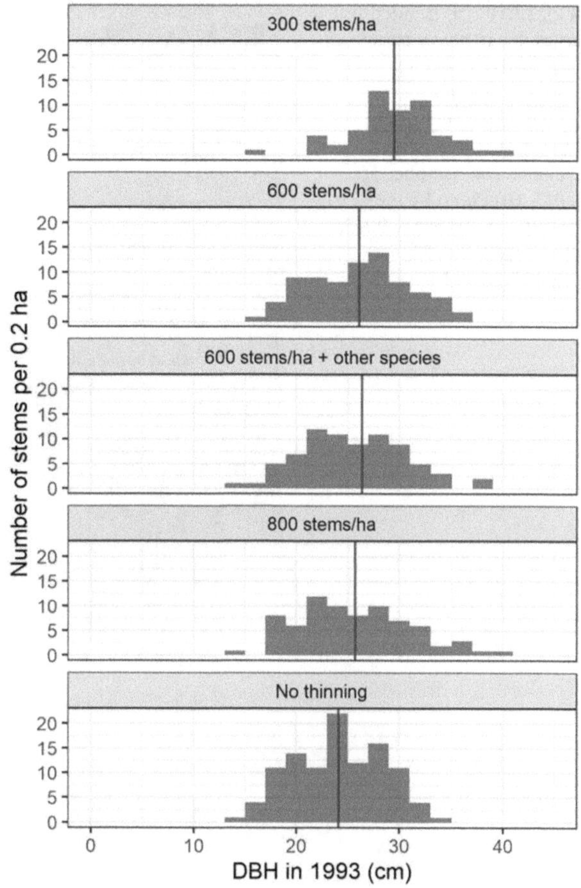

Fig. 13.9 DBH distribution in a thinning experiment plot (Plot No. 5401) in 1993. Thinning treatments were carried out in 1955. The stems of the monarch birch were thinned to the specified density. The stems of the other species were cut, except for those in the "600 stems/ha + other species" treatment. Red lines indicate the mean DBH. Note that the average DBH in the stand with 300 stems/ha was clearly larger than that in the no-thinning stand. (Redrawn after an unpublished article by H. Inukai et al.)

before the leaf-shedding of deciduous broadleaves can lead to further accumulation of snowpack, which may have exacerbated the damage (Fig. 13.10b). Although the total amount of damage was not reported, several experimental plots experienced severe damages at the 2013 wet snow. For instance, in a larch plantation in experimental Plot No. 1051, out of 138 trees, 5 trees were bent, 27 trees were snapped, and 15 trees were downed. In another *Larix* plantation in experimental Plot No. 1050, out of 465 trees, 5 trees were bent, 30 trees were snapped, and 53 trees were downed or learned. In *B. platyphylla* var. *japonica* plantation in experimental plot 3008, out of 37 *Betula* trees, 16 trees were bending, 5 trees were snapped, and 13 trees were downed.

Controlling stem density is important for resistance to snow damage. In the density control experimental plots of *B. platyphylla* var. *japonica* (Plot No. 3029), stem density steeply decreased from 613 stems ha^{-1} to 193 stems ha^{-1} between 2012 and

Fig. 13.10 (**a**) Bended trees due to heavy snow load and (**b**) heavily damaged *Betula platyphylla* stand in 2013. ((**a**) Photo by Sunisa Sanguansub, October 16, 2013, (**b**) Photo by the UTHF, November 7, 2013, licensed under CC BY-NC-ND 4.0)

2020 in the highest density plot, mostly due to the snow damage in 2013 (initially 4000 stems ha^{-1}) (Fig. 10.12 in Chap. 10). In contrast, in the lowest-density plot (initially 2000 stems ha^{-1}), stem density decreased more gradually, from 753 stems ha^{-1} to 507 stems ha^{-1}. This suggests that thin stems in stands with high stem density are more vulnerable to snow packing.

13.4 Other Disturbances

Heavy rains, often associated with typhoons, can cause landslide and floods. However, large-scale landslides and floods have been rare in the UTHF, although heavy rains can occasionally destroy forest roads (refer to Chap. 9).

Although insect outbreak is also one of causes of a natural disturbance in the UTHF (refer to Chap. 14), most of them were not lethal for trees and did not cause stand-replacing disturbances. An exception is the spruce bark beetle *Ips typographus japonicas*, which usually attacks weakened *Picea* trees but occasionally attacks healthy trees, especially after wind disturbances or logging (refer to also Chap. 12).

13.5 Remarks

Monoculture plantations are considered vulnerable to disturbances such as windstorms (Morimoto et al. 2019), insect outbreaks (Jactel et al. 2005), and snow damage. The frequency and severity of biotic and abiotic disturbances are expected to increase due to ongoing climate change. Natural mixed conifer-broadleaf forests, which are uneven-aged multispecies stands, are believed to have greater resistance and resilience to such disturbances. UTHF has shifted from typical monoculture plantations to low-density polyculture plantations, aiming to restore mixed conifer-broadleaf forests and to improve the resistance and resilience to disturbances (refer to Chap. 10).

References

Asama M, Takahashi Y, Okamura K, Yamamoto H (1996) Restorative afforestation of stands damaged by the typhoon in 1981: growth of planted *Abies sachalinensis* and natural regeneration on after site preparation. Trans Mtg Hokkaido Br Jpn For Soc 44:61–63. (in Japanese)

Duperat M, Gardiner B, Ruel J-C (2020) Wind and snow loading of balsam fir during a Canadian winter: a pioneer study. Forests 11:1089. https://doi.org/10.3390/f11101089

Gardiner B (2021) Wind damage to forests and trees: a review with an emphasis on planted and managed forests. J For Res 26:248–266. https://doi.org/10.1080/13416979.2021.1940665

Hirata T, Maezawa K (1956) Wind damages in the Tokyo university Forest in Hokkaido. Misc Info Univ Tokyo Forests 11:1–8

Hotta W, Morimoto J, Haga C, Suzuki SN, Inoue T, Matsui T, Owari T, Shibata H, Nakamura F (2019) Long-term cumulative impacts of windthrow and subsequent management on tree species composition and aboveground biomass: a simulation study considering regeneration on downed logs. For Ecol Manag 502:119728

Hotta W, Morimoto J, Haga C, Suzuki SN, Inoue T, Matsui T, Owari T, Shibata H, Nakamura F (2021) Long-term cumulative impacts of windthrow and subsequent management on tree species composition and aboveground biomass: a simulation study considering regeneration on downed logs. For Ecol Manag 502:119728. https://doi.org/10.1016/J.FORECO.2021.119728

Hotta W, Haga C, Morimoto J, Suzuki SN, Matsui T, Owari T, Shibata H, Nakamura F (2023) Leaving disturbance legacies conserves boreal conifers and maximizes net CO_2 absorption under climate change and more frequent and larger windthrow regimes. Landsc Ecol 38:1785–1805. https://doi.org/10.1007/s10980-023-01680-4

Jactel H, Brockerhoff E, Duelli P (2005) A test of biodiversity-stability theory: meta-analysis of tree species diversity effects on insect pest infections, and re-examination of responsible factors. In: Scherer-Lorenzen M, Körner C, Schulze E-D (eds) Forest diversity and function, Ecological studies, vol 176. Springer, Berlin, pp 235–262

Kikuzawa K (1989) A thinning effect in a young birch stand. Trans Mtg Hokkaido Br Jpn For Soc 37:49–51. (in Japanese)

Kisanuki H, Kimura N, Kurahashi A, Kaji M (1996) Regeneration process of the forest after typhoon damage in the mixed forest with coniferous and broad leaved species. — stand condition where 14 years passed after the typhoon damage. Mtg Hokkaido Br Jpn For Soc 44:58–60

Li J, Morimoto J, Hotta W et al (2023) The 30-year impact of postwindthrow management on the forest regeneration process in northern Japan. Landsc Ecol Eng 19:227–242. https://doi.org/10.1007/s11355-023-00539-9

Matsui M, Okamura K, Okahira T, Goto S (2008) Possibility of establishing mixed forests by combining Picea jezoensis planting and natural regeneration in low elevation zone. Japanese J Forest Environ 50:35–40. (in Japanese)

Miura S, Matsumoto Y, Mashimo Y (1984) Analyses of the wind damage caused by typhoon 15 in 1981 at Tokyo university Forest in Hokkaido. Bull Tokyo Univ For 73:139–160

Miyoshi C, Takahashi H, Kozuka C, Naniwa A, Sakai R, Ito K, Forestry Techinical Staff (2014) Thinning of secondary birch stands regenerated after scarification in the Teshio experimental Forest. Tech Rep Boreal Forest Conserv Hokkaido Univ Forests 31:1–8. (in Japanese)

Morimoto J, Nakagawa K, Takano KT, Aiba M, Oguro M, Furukawa Y, Mishima Y, Ogawa K, Ito R, Takemi T, Nakamura F, Peterson C (2019) Comparison of vulnerability to catastrophic wind between Abies plantation forests and natural mixed forests in northern Japan. Forestry 92:436–443. https://doi.org/10.1093/forestry/cpy045

Nagel TA, Mikac S, Dolinar M, Klopcic M, Keren S, Svoboda M, Diaci J, Boncina A, Paulic V (2017) The natural disturbance regime in forests of the Dinaric Mountains: a synthesis of evidence. For Ecol Manag 388:29–42. https://doi.org/10.1016/J.FORECO.2016.07.047

Okamura K, Takahashi Y, Ota S, Kawahara S, Sima T, Hirata M, Yamamoto H (1995) Recovery of stands damaged by the typhoon in 1981: growth of the artificial and natural regenerations for 11 years. Trans Mtg Hokkaido Br Jpn For Soc 43:189–191. (in Japanese)

Okamura K, Shima T, Mastui M, Yamamoto K, Goto S (2006) Effectiveness of lined ground preparation for reforestation in damaged areas by a typhoon (II) growth of 13-years after early salvage cutting. Trans Mtg Hokkaido Br Jpn For Soc 54:36–38. (in Japanese)

Sano Y, Shibuya M (2015) Thinning effect on tree growth in secondary *Betula ermanii* stands on scarified sites. J Jpn For Soc 97:25–29. (in Japanese with English summary)

Shibano S, Ohta S, Ito T, Watanabe S (1990) Topographical analyses of wind damaged by typhoon-15 in 1981: a case of the Tokyo university Forest in Hokkaido. Trans Mtg Hokkaido Br Jpn For Soc 38:212–214

Shimizume M, Shima T, Iguchi K, Okahira T (2002) Validity of ground clearance for the reforestation in damaged areas by the typhoon: the growth and crown damage of 15-yr *Abies sachalinensis* planted there. Trans Mtg Hokkaido Br Jpn For Soc 50:32–34

Suvanto S, Lehtonen A, Nevalainen S, Lehtonen I, Viiri H, Strandström M, Peltoniemi M (2021) Mapping the probability of forest snow disturbances in Finland. PLoS One 16(7):e0254876. https://doi.org/10.1371/journal.pone.0254876

Suzuki S (2020) Tokyo daigaku Hokkaido ensyurin no tennennrinn keizoku kansoku no gaiyou [outline of permanent plots in natural forests in the University of Tokyo Hokkaido Forest (translated by authors from Japanese)]. Northern Foresty Japan 71(3):16–21

Suzuki SN, Tsunoda T, Nishimura N, Morimoto J, Suzuki J-I (2019) Dead wood offsets the reduced live wood carbon stock in forests over 50 years after a stand-replacing wind disturbance. For Ecol Manag 432:94–101

Takada N, Inukai H, Watanabe S (1988) Evaluation of *Betula maximowicziana* excellent tree in secondary stands. Trans Mtg Hokkaido Br Jpn For Soc 37:241–243

Tokyo University Forest in Hokkaido (1973) The 75th history of the Tokyo university Forest in Hokkaido. Tokyo University Forests, p 116. (in Japanese)

Tokyo University Forests (1950) The 50th history of the Tokyo University Forest in Hokkaido, p 61. (in Japanese)

Uchiyama K, Goto S, Tsuda Y, Takahashi Y, Ide Y (2006) Genetic diversity and genetic structure of adult and buried seed populations of Betula maximowicziana in mixed and postfire stands. For Ecol Manag 237:119–126

Watanabe S, Shibata S, Kawahara S, Shibano S, Kurahashi A, Satoo Y, Anazawa C, Takada N, Takahashi Y (1990) A memoir on the actual situation of the forest wind-damaged by the typhoon no. 15 in 1981 in the Tokyo university Forest in Hokkaido. Misc Info Univ Tokyo Forests 27:79–221

Chapter 14
Biological Threats: Insect Pests, Diseases, and Mammal Damage

Dai Fukui, Naoto Kamata, and Daisuke Sakaue

Abstract Damage to forests caused by insect pests, diseases, and wildlife leads not only to the loss of forest resources but also to a decline in the ecosystem functions of forests. At the UTHF, damage to forests caused by insect pests, diseases, and wildlife has been a problem that must be solved throughout its history. This chapter introduces the forest damage caused by these events that has occurred at the UTHF to date, as well as examples of research into the mechanisms behind their occurrence.

Keywords Defoliator · Wood borer · Fungal disease · Wood decay · Snow mold · Rodent · Sika deer

14.1 Insect Pests

Insects in some feeding guilds, such as foliage feeders (defoliating insects) and wood borers (*sensu lato*), including phloem feeders, have caused serious damage to natural forest stands and/or plantations in the UTHF. Severe defoliations were sometimes followed by secondary attacks by the wood borers (*sensu lato*).

D. Fukui (✉)
The University of Tokyo Fuji Iyashinomori Woodland Study Center, Graduate School of Agricultural and Life Sciences, The University of Tokyo, Yamanakako, Japan
e-mail: fukuidai@uf.a.u-tokyo.ac.jp

N. Kamata
The University of Tokyo Chiba Forest, Graduate School of Agricultural and Life Sciences, The University of Tokyo, Kamogawa, Japan
e-mail: kamatan@uf.a.u-tokyo.ac.jp

D. Sakaue
The University of Tokyo Chichibu Forest, Graduate School of Agricultural and Life Sciences, The University of Tokyo, Chichibu, Japan
e-mail: sakaue@uf.a.u-tokyo.ac.jp

T. Owari et al. (eds.), *Integrative Forest Management and Silviculture*,
Ecological Research Monographs, https://doi.org/10.1007/978-981-95-1017-7_14

14.1.1 Defoliating Insects

Saturn Moth, *Caligula japonica* (Lepidoptera: Saturniidae)
Caligula japonica (Lepidoptera: Saturniidae: Fig. 14.1) is endemic to Japan and is widely distributed from Kyushu to Hokkaido. Populations of *C. japonica* sometimes reach outbreak levels and cause severe defoliation of the host trees. The caterpillar is polyphagous, although its host preference differs among regions. In Hokkaido, Monarch birch, *Betula maximowicziana,* is preferred.

In the UTHF, population outbreaks of *C. japonica* have been recorded since the 1990s. Several years of severe defoliation sometimes causes branch dieback and tree mortality in *B. maximowicziana.* The population outbreak from 2009 to 2012 was most severe. In 2010 and 2011, large areas of *B. maximowicziana* were severely defoliated. Some trees died after several years of severe defoliation. Some surviving individuals suffer secondary attacks by ambrosia beetles, which greatly reduce market values (Kamata et al. 2014). Branch dieback and damage caused by wood boring insects are serious problems for the production of high-value timber from *B. maximowicziana.* Therefore, appropriate management, such as thinning of weakened trees, is important soon after successive years of severe insect defoliation.

According to adult monitoring conducted by Iguchi (2023) at streetlights in the town area of Furano from late August to early September, a few individuals or fewer were recorded per streetlight in 2018–2020, and approximately 30 individuals were recorded in 2021, while 500–1000 individuals were recorded in 2022 (Fig. 14.3). Although the number of individuals around streetlights decreased to around 10 individuals per streetlight in 2023, severe insect defoliation on *B. maximowicziana* was found in the UTHF and its surrounding areas. Like monoculture plantations, secondary forest stands of *B. maximowicziana*, which regenerate after forest fires, are less diverse both in terms of plant species composition and tree age. As predicted by the diversity-stability hypothesis (Elton 1958) and the resource concentration hypothesis (Root 1973), defoliation by the *C. japonica* population tended to be

Fig. 14.1 Saturn moth, *Caligula japonica* (Lepidoptera: Saturniidae) (Photo by Naoto Kamata) (**a**) and its larvae. (Photo by Asami Sanyoshi, used with permission from the UTHF, licensed under CC BY-NC-ND 4.0) (**b**)

Fig. 14.2 Natural secondary forest at Nishinosawa dominated by *Betula maximowcziziana* defoliated by *Caligula japonica* in 2012, which regenerated after forest fire in early twentieth century. (Photo by Naoto Kamata, August 8, 2012)

Fig. 14.3 *Caligula japonica* moths attracted to a streetlight in Furano City in 2022. (Photo by Kazunobu Iguchi, September 2, 2022, licensed under CC BY-NC-ND 4.0)

more severe in secondary forest stands of *B. maximowicziana* than in old-growth natural forests (Fig. 14.2).

***Betula maximowicziana* Defoliation by a Mixture of *Operophtera brumata* and *Erannis golda* (Lepidoptera: Geometridae)**

In Hokkaido, severe defoliations by a mixture of two species belonging to the family Geometridae, *Operophtera brumata* and *Erannis golda*, has been observed at approximately 10-year intervals on *Alnus* spp., *Betula* spp., *Prunus* spp., *Acer* spp., *Quercus crispula*, and *Tillia japonica* (FRD HRO 1995). At Nishinosawa in the UTHF, an approximately 100-ha of secondary forest stand dominated by *B. maximowicziana* that regenerated after forest fires suffered severe defoliation by the two species in June 1995 (Iguchi 2023). Within this stand, thirty trees with severe defoliation died in the following year, although no conspicuous defoliation was observed in 1996. However, in May 1996, trees that had suffered severe defoliation in the previous year but survived started to experience secondary attacks by *Heteroborips seriatus* (Coleoptera: Curculionidae: Scolytinae) and other ambrosia beetles. Unusual for ambrosia beetles, *H. seriatus* is known to bore galleries in inner bark as well as wood (Fig. 14.4) and carry an unidentified species belonging to the genus *Ophiostoma*, a blue stain fungus that kills sapwood cells. In the same secondary forest stand at Nishinosawa, approximately 100 trees (with a timber volume of 435 m^3) were killed by the secondary attacks, and approximately 5000 trees (with a timber volume of 2500 m^3) that were still alive but exhibited severe branch dieback were harvested in 1996. Trees with abundant foliage were not logged. However,

Fig. 14.4 Ambrosia beetle, *Heteroborips seriatus* (Coleoptera: Curclionidae) adults boring the inner bark of *Betula maximowicziana*. (Photo by Naoto Kamata)

discolorations by fungi can develop in the wood of these surviving trees if they are subjected to secondary attacks by ambrosia beetles. After that, no damage caused by Geometridae was confirmed.

Larch Sawfly, Pristiphora erichsoni (Hymenoptera: Tenthredinidae)

The larch sawfly, *Pristiphora erichsoni* (Hymenoptera: Tenthredinidae), is a cosmopolitan species that is widely distributed where larches, *Larix* spp. grow. It is well known that population outbreaks of *P. erichsoni* persist for extended periods, even with changes in location. Larches are not native to Hokkaido but were introduced in the twentieth century; thus, *P. erichsoni* is also not native to Hokkaido. The first outbreak in Hokkaido was recorded in 1932–1933 in the Niikappu, Hidaka Region (FRD HRO 2010). The outbreak that started in 1977 lasted for 11 years. The latest outbreak started in 1995 on the Oshima peninsula and moved eastward. The outbreaks continued until 2022 and ended in the eastern part of Hokkaido, while mild insect defoliation was observed in central Hokkaido in 2023 (FRD HRO 2010). Severe defoliation caused by the most recent outbreak at the UTHF was first recognized in 2009 and continued through 2016 (Iguchi 2023). Conspicuous defoliation was observed in any of the stands every year. The Japanese larch, *Larix kaempferi,* was most severely defoliated, followed by the Korean larch, *L. olgensis*. Kuril larch, *L. gmelinii,* was least preferred. The mortality rate caused by entomopathogenic microorganisms is generally low. Predation by small mammals and mortality caused by parasitoids, such as tachinid flies and ichneumonid wasps, sometimes increase but cannot regulate population outbreaks, probably because small mammals prey on cocoons parasitized by parasitoids. The populations of small mammals decreased in summer, because *P. erichsoni* cocoons are seasonal food. Nitrogen deficit due to successive years of defoliation deteriorated foliage quality as food (Pinkantayong et al. 2015), which was likely the cause of termination of the outbreak. Larch trees seldom die even in the case that they suffer successive years of severe defoliation. However, secondary mass attacks by the larch bark beetle, *Ips cembrae* (Coleoptera: Curclionidae: Scolytinae), cause a clump of tree mortality in some stands (Figs. 14.5 and 14.6).

Hesperiid Butterfly, Bibasis aquiline (Lepidoptera, Hesperiidae)

Bibasis aquilina is a Palearctic butterfly distributed in Japan, China, and Amur. The larva feeds on *Kalopanax septemlobus* foliage and occasionally causes severe defoliation. Since the larva creates a simple nest by folding a leaf, feeding damage caused by this species is easily identifiable. An outbreak of this species in the UTHF occurred in 1980. Since severe defoliation caused by this species had already been shown to lead to branch dieback and tree mortality within a few years, as observed during local damage in the 1970s, 712 trees (2130 m^3 in volume) that lost more than two-thirds of their foliage were harvested during the outbreak in 1980 due to their high likelihood of dying (Shibata and Iguchi 1981). At that time, a large influx of large-diameter *K. septemlobus* timber into the timber market in Asahikawa resulted in a decline in its value. A point of caution is that, even if damaged tree does not die, wood decay may progress to the trunk from the diebacked branches triggered by secondary infestations from wood-boring insects such as longhorn beetles, leading

Fig. 14.5 Larch sawfly, *Pristiphora erichsoni (Hymenoptera: Tenthredinidae)*, depositing an egg inside a long shoot tissue of the Japanese (**a**), eggs individually laid inside a long shoot tissue (**b**), larvae feeding on short shoots in a group (**c**), cocoons mostly preyed by small mammal predators (**d**). (Photos by Naoto Kamata)

Fig. 14.6 Tree mortality in the larch plantation near the UTHF caused by successive years of severe defoliation by the larch sawfly, *Pristiphora erichsoni.* (Photos by Naoto Kamata, August 19, 2013)

to significant deterioration of timber quality. No large-scale population outbreaks had been observed since 1980, although individual trees have sometimes showed signs of the insect defoliation. Conspicuous defoliation by *K. septemlobus* was observed throughout the UTHF in 2019, marking the first large-scale outbreak since 1980.

Linden Leaf Mining Sawfly, Parna kamijoi (Hymenoptera, Tenthredinidae)

Parna kamijoi (Hymenoptera, Tenthredinidae) is a leaf mining sawfly on lindens, *Tillia* spp., one of a dominant broadleaf genus in the UTHF (Fig. 14.7). Both *T.*

Fig. 14.7 *Parna kamijoi* (Hymenoptera: Tenthredinidae) larva mining *Tillia japonica* leaf. (Photo by Kazunobu Iguchi, licensed under CC BY-NC-ND 4.0)

maximowicziana and *T. japonica* are the host plants in the UTHF. The feeding period of the larvae is approximately one month, from mid-May to mid-June. In the UTHF, six outbreaks were observed at 3-year intervals between 1988 and 2003. Like those of periodic cicadas, the mechanisms of 3-year periodic outbreaks depend on the specific cohort. Approximately 95% of individuals complete their life cycle in three years due to prolonged pupal diapause, although the remainder emerge outside the 3-year periodicity. The population consists of three cohorts, with one cohort having a much greater number of individuals, resulting in noticeable outbreaks every three years. (Iguchi et al. 1998, 2000; Iguchi and Ozaki 2006).

When the defoliation by *P. kamijoi* was first recognized, it did not reach the level of population outbreak. However, the defoliation area continued to expand every 3 years, and the defoliation intensity also continued to increase. While no tree mortality following defoliation has been observed, branch diebacks have become noticeable. Conspicuous defoliations were also recorded in 2014, 2015, 2017, 2020, and 2022, with the 2017 and 2020 events being particularly widespread and severe. It is likely that appropriate measures will be necessary in the near future. It is also necessary to determine the impact of defoliation by *P. kamijoi* on individual trees.

Other Defoliators

Population outbreaks of some forest defoliators are known to be related to stand age (Kamata 2002). Those of the following species possibly tend to occur in plantations at a certain status, because no outbreaks have been recorded for a long period.

The pine caterpillar, *Dendrolimus spectabilis* (Lepidoptera: Lasiocampidae), feeds on needles of species belonging to the genus *Pinus*. Generally, it is known that outbreaks tend to occur in young plantations of *Pinus* spp. However, in Hokkaido, there have been reports of feeding damage in garden trees, but few outbreaks have been recorded in plantations. The outbreak in the UTHF occurred in 1985 at a 68-year-old plantation of eastern white pine, *Pinus strobus,* planted in 1917 (Arisawa et al. 1986). The most severely damaged stands were found in the plots without thinning at the thinning experiment site (area: approximately 1 ha). Because the individual crowns of *P. strobus* were poorly developed in the plot and the foliage biomass before the outbreaks was small, trees growing there were probably less vigorous so that most of the individuals died in the following year of severe defoliation were found in the plots without thinning (Arisawa et al. 1986). Since the severely damaged area was adjacent to an exhibition forest, intensive pest control was carried out by the UTHF. First, in the fall of 1985, the trunk was wrapped in a straw mat, and the larvae that came down on the trunk surface for overwintering were captured by the mat and killed. In the following spring, insecticide was sprayed from the ground to the entire canopy of the pine trees around the damaged area, including the exhibition forest. The pine trees in the severely damaged areas (area: approximately 4 ha) were clearly cut. In addition, the larvae were mostly parasitized by parasitic wasps and parasitic flies after overwintering (Arisawa et al. 1987). As a result, the outbreak ended in one year. Thereafter, no outbreaks have been recognized in the UTHF.

Population outbreaks of the red-headed spruce web-spinning sawfly, *Cephalcia isshikii,* were recognized in 1944 in *Picea abies* stands planted in the 1910s (Nitto and Akui 1957). According to the intensive survey 1953, defoliation tended to be severe in fast-growing plantations, where the canopy was closed with dense foliage and understory vegetation was poor. In contrast, defoliation was mild in plantations with dense dwarf-type bamboo bushes. Twigs of severely defoliated trees were sometimes infested by *Pityophthorus* (=*Cladoborus*) *arakii* (Coleopteara: Curclionidae: Scolytinae).

14.1.2 Wood Borers

Eight-Toothed Spruce Bark Beetle, Ips typographus japonicus (Coleoptera: Curclionidae)

Ips typographus japonicus is a bark beetle belonging to the subfamily Scolytinae and is distributed on the Honshu and Hokkaido Islands. The base subspecies, *Ips typographus,* is widely distributed on the Eurasian continent. These are known as the most devastating pests of *Picea* spp. from Europe to Asia. These are basically secondary insects that attack weakened trees, broken branches, and felled trees. However, once the population density increases, healthy-looking trees are attacked and the host trees are killed by mass attacks. These fungi have symbiotic relationships with blue stain fungi, such as *Ophiostoma* spp. and *Ceratocystis* spp. Mass inoculation by blue stain fungi by mass attacks is a direct cause of tree mortality. In particular, the abundance of *I. typographus japonicus* increased greatly after large-scale forest disturbances, such as a typhoon; a great number of Ezo spruce, *Picea jezoensis,* trees were killed by mass attacks. After Typhoon Thad in 1981 (T8115, refer to Chap. 13), approximately 30,000 m^3 of *Picea* trees, including *P. jezoensis* as well as *P. glehnii,* were infested by *I. typographus japonicus* and harvested, accounting for 68% of the total insect damage at that time (Watanabe et al. 1990). Ordinary forestry practices in the UTHF also cause population outbreaks of *I. typographus japonicas* on a local scale and kill healthy-looking trees. When selection cutting is carried out in a natural forest stand dominated by *P. jezoensis,* tree mortality caused

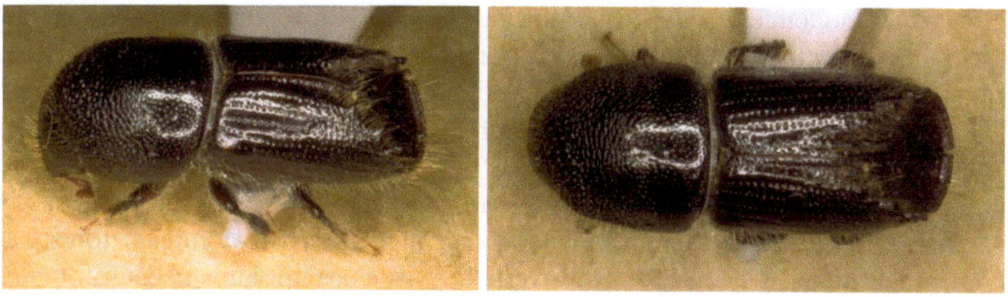

Fig. 14.8 Eight-toothed spruce bark beetle, *Ips typographus japonicus (Coleoptera: Curclionidae)*. (Photos by Marek Turčáni, licensed under CC BY-NC-ND 4.0)

by insect mass attacks occurs approximately three years later. There is no relationship between mortality and tree age or the absence/presence of wood decay (Iguchi et al. 1999), but mortality is greater around logging landings (Iguchi et al. 1993a, b, refer to also Chap. 12), probably due to greater disturbance and/or greater amounts of volatile emissions from *P. jezoensis* resulting from timber construction. Preparing *P. jezoensis* trees around the timber yards is effective at suppressing the total amount of tree mortality (Iguchi et al. 1993a, b). This species is presumed to cause similar damage to Sakhalin spruce, *Picea glehnii*. In the UTHF, many *P. glehnii* plantations will reach the age for thinning, so it is necessary to treat them carefully to prevent *I. typographus japonicus* outbreaks (Fig. 14.8).

14.2 Forest Diseases and Their Causal Fungi

14.2.1 Overview

Generally, severe forest damage caused by disease rarely occurs in natural forests, especially those with old growth, where multiple tree species of varying ages coexist. In the UTHF, most of the natural forests are mixed forests, even in subalpine coniferous forests where *Picea jezoensis* is usually mixed with *Abies sachalinensis* and broad-leaved trees such as *Betula ermanii*, *Sorbus commixta* and *Acer ukurunduense* (refer to Chap. 5). Thus, monospecific or nearly monospecific forest stands are limited to (sub)alpine *Pinus pumila* forests above the tree (forest) line, secondary forests after fires dominated by *Betula maximowicziana*, and plantations. Therefore, important diseases that cause severe damage usually occur on plantations.

In the UTHF, while artificial planting was conducted following the forest fires of the 1910s and the typhoon of 1954 as part of forest restoration efforts (refer to Chap. 10), large-scale and systematic artificial planting commenced in earnest and on a continuous basis after 1958, coinciding with the adoption of the Stand-based Silvicultural Management System. This aimed to improve and transform the forest stand structure into a high-growth structure. During the 1960s–1970s, certain diseases occasionally caused serious damage to juvenile plantations. However, these diseases are no longer a significant issue, as the area of juvenile stands with monospecific and homogenous structures has declined in response to the shift in forest stand improvement policies. At present, artificial planting is conducted to assist the natural regeneration of conifers, with high-density, single-species planting being deliberately avoided (refer to Chap. 10).

In natural forests, one of the few diseases that can cause sudden outbreaks is vascular wilt caused by ophiostomatoid fungi in *P. jezoensis*-dominant forests at higher elevations (Fig. 14.9). However, the virulence of known causal fungi, *Endoconidiophora polonica* (≡ *Ceratocystis polonica*) and *Ceratocystis penicillata* (= *Ophiostoma penicillatum*), is generally weak and mass inoculation is needed even though these fungi have the ability to kill host trees (Yamaoka et al. 2000). Therefore, the outbreak of the disease is largely driven by the dynamics and abundance of the vector bark beetle, *Ips typographus japonicus* (refer to Sect. 14.1.2). The occurrence of damage is usually limited to a small area except after large disturbances such as typhoons. For countermeasures, deforestation in neighboring

Fig. 14.9 A dead *Picea jezoensis* tree infested by a bark beetle *Ips typographus japonicus* accompanying ophiostomatoid fungi. (Photo by Daisuke Sakaue, May 17, 2011)

areas over the years should be avoided, *P. jezoensis* trees around logging landings should be removed, and the installation of pheromone traps is recommended, all of which aim to avoid increasing the number of bark beetles (Nakayama et al. 1991; Iguchi et al. 1993a, b). This kind of damage could also occur in young plantations of *P. glehnii* after cleaning cuts and thinning (Hara and Hayashi 2002), and in plantations of *Larix* spp. In the latter case, the bark beetle, *Ips subelongatus*, which accompanies the fungus *Endoconidiophora laricicola* (≡ *Ceratocystis laricicola*), attacks, and the damage is sometimes combined with Armillaria root rot (Tokuda and Onodera 2018).

14.2.2 Fungal Flora That Cause Foliar, Stem, and Root Diseases

After the outbreak of needle rust in juvenile (less than 10 years of stand age) *Pinus strobus* plantations in 1958, an energetic survey of forest diseases that occur in the UTHF was conducted during 1960–1972. The results were summarized by Saho and Takahashi (1973), who listed 65 species of fungi pathogenic to forest trees that were both native and introduced to Hokkaido. Following this, Takahashi (1979) conducted an intensive survey during 1972–1974 in experimental plantations along an elevational gradient (530–1100 m a.s.l., Plot Nos. 3020–3023, refer to Sect. 10.2.2.4). At these experimental sites, the native conifers (*Abies sachalinensis*, *Picea jezoensis*, and *P. glehnii*) and introduced *Pinus* (*P. strobus*, *P. sylvestris*, and *P. banksiana*) and *Larix* (*L. kaempferi*, *L. gmelinii* varieties and hybrids, *L. sibirica*, and *L. laricina*) species were planted. The survey also included *A. sachalinensis*, *P. jezoensis*, *P. glehnii*, and *Pinus pumila* in the nearby natural forest. Through his survey, Takahashi (1979) reported 66 species of pathogenic fungi and revealed the flora and their variation with altitude, together with the host range of each fungus.

Here, the pathogenic fungi that cause foliar, stem and root diseases are summarized in Table 14.1 based on Saho and Takahashi (1973), Takahashi (1979), and a few other reports. Species identification follows the reexamination by Kobayashi (2007), with scientific names aligned with the GBIF—the Global Biodiversity Information Facility database (https://www.gbif.org/species/). Disease names are sourced from the "Common name database of plant diseases in Japan" by the Genebank Project, NARO, Japan (https://www.gene.affrc.go.jp/databases-micro_pl_diseases.php).

A total of 75 species are recognized in Table 14.1. Among them, five species were newly described and 30 species were reported from Japan for the first time. Twenty-seven species were found on both native and introduced tree species, while 16 and 32 species were found exclusively on native and introduced hosts, respectively. As a side note, Takahashi (1979) reported an additional seven unidentified pathogenic fungal species. It should be noted that Table 14.1 includes species

Table 14.1 Pathogenic fungi causing foliar, stem, and root diseases observed in the University of Tokyo Hokkaido Forest

Scientific name	Disease name	Host			Family	Order
		Genus	Native	Introduced		
Phaeocryptopus nudus	Adelopus needle cast	*Abies, Picea, Pinus*	○		Capnodiaceae	Capnodiales
Racodium therryanum	Racodium snow blight	*Abies, Picea*	○		Racodiaceae	Capnodiales
[b]*Hysterium acuminatum*	(Not proposed)	*Picea*		○	Hystiriaceae	Hysteriales
Dothistroma septosporum	Dothistroma needle blight	*Pinus*		○	Mycosphaerellaceae	Mycosphaerellales
Septoria pini-pumilae	Septoria needle blight	*Pinus*	○	○	Mycosphaerellaceae	Mycosphaerellales
[b]*Lophium mytilinum*	Lophium twig blight	*Abies, Larix, Picea, Pinus*	○	○	Mytilinidiaceae	Mytilinidiales
Curreya pityophila	(not proposed)	*Pinus*	○	○	Cucurbitariaceae	Pleosporales
Rhizosphaera kalkhoffii	Rhizoshaera needle blight	*Pinus*		○	Venturiaceae	Venturiales
[a]*Macrophoma yamabeana*	(Not proposed)	*Picea*	○	○	Botryosphaeriaceae	Botryosphaeriales
Sphaeropsis sapinea	Diplodia needle and twig blight	*Pinus*		○	Botryosphaeriaceae	Botryosphaeriales
Guignardia laricina	shoot blight	*Larix*		○	Phyllostictaceae	Botryosphaeriales
[b]*Cyclaneusma niveum*	Naemacyclus needle cast	*Pinus*		○	Marthamycetaceae	Chaetomellales
[b]*Cenangium acuum*	Cenangium needle blight	*Pinus*		○	Cenangiaceae	Helotiales
Cenangium ferruginosum	Cenangium twig blight	*Pinus*		○	Cenangiaceae	Helotiales
Dermea abietina (doughtful)	(Gelatinosporium twig blight)	*Pinus*	○	○	Dermateaceae	Helotiales
[b]*Dermea balsamea*	Dermea canker	*Abies*	○	○	Dermateaceae	Helotiales
[b]*Dermea pinicola*	Dermea canker	*Pinus*	○	○	Dermateaceae	Helotiales
[b]*Pezicula livida*	(not proposed)	*Larix, Pinus*		○	Dermateaceae	Helotiales
[b]*Ascocalyx abietis*	Ascocalyx canker	*Abies, Larix*	○	○	Godroniaceae	Helotiales
[b]*Godronia multispora*	Godronia canker	*Betula*		○	Godroniaceae	Helotiales

Scientific name	Disease name	Host			Family	Order
		Genus	Native	Introduced		
[b]*Gremmeniella abietina*	Scleroderris canker	*Abies, Pinus*	○	○	Godroniaceae	Helotiales
[b]*Gremmeniella laricina*	Encoeliopsis canker	*Larix*		○	Godroniaceae	Helotiales
[b]*Gremmenia abietis*	Phacidium snow blight	*Abies, Picea, Pinus*	○	○	Hamatocanthosxyphaceae	Helotiales
[b]*Dasyscyphus pini*	Lachnellula canker	*Pinus*	○	○	Hyaloscyphaceae	Helotiales
[b]*Lachnellula arida*	Lachnellula canker	*Larix*		○	Lachnaceae	Helotiales
Lachnellula calyciformis	Lachnellula canker	*Abies, Larix, Picea, Pinus*	○	○	Lachnaceae	Helotiales
[b]*Lachnellula microspora*	Lachnellula canker	*Abies, Pinus*	○		Lachnaceae	Helotiales
[b]*Lachnellula subtilissima*	Lachnellula canker	*Pinus*	○	○	Lachnaceae	Helotiales
[b]*Lachnellula suecica*	Lachnellula canker	*Larix, Picea, Pinus*	○	○	Lachnaceae	Helotiales
Lachnellula willkommii	Lachnellula canker	*Larix*		○	Lachnaceae	Helotiales
Lachnellula sp.	(Lachnellula canker)	*Picea*	○		Lachnaceae	Helotiales
Lachnellula sp.	(Lachnellula canker)	*Abies*	○		Lachnaceae	Helotiales
[b]*Tympanis hypopodia*	Tympanis canker	*Pinus*	○	○	Tympanidaceae	Helotiales
[b]*Tympanis laricina*	Tympanis canker	*Abies, Larix*	○	○	Tympanidaceae	Helotiales
[b]*Tympanis spermatiospora*	(not proposed)	*Populus*		○	Tympanidaceae	Helotiales
[b]*Tympanis truncatula*	Tympanis canker	*Pinus, Pseudotsuga*		○	Tympanidaceae	Helotiales
[b]*Tympanis tsugae*	Tympanis canker	*Picea*	○		Tympanidaceae	Helotiales
[a]*Waltonia pinicola*	Waltonia twig blight	*Pinus*		○	–	Helotiales
[b]*Phacidium abietinum*	(not proposed)	*Picea*		○	Phacidiaceae	Phacidiales
[b]*Discocainia treleasei*	Atropellis canker	*Picea*	○		Rhytismataceae	Rhytismatales

(continued)

Table 14.1 (continued)

Scientific name	Disease name	Host			Family	Order
		Genus	Native	Introduced		
Hypodermella sp.	(needle cast)	*Pinus*	○		Rhytismataceae	Rhytismatales
Lirula nervisequa	(not proposed)	*Abies, Picea*	○		Rhytismataceae	Rhytismatales
Lophodermium abietis	needle cast	*Abies*	○		Rhytismataceae	Rhytismatales
[b]*Lophodermium piceae*	needle cast	*Abies, Picea*	○		Rhytismataceae	Rhytismatales
Lophodermium pinastri	needle cast	*Pinus*	○	○	Rhytismataceae	Rhytismatales
[b]*Tryblidiopsis pinastri*	Tryblidiopsis twig blight	*Larix, Picea, Pinus, Pseudotsuga*	○	○	Rhytismataceae	Rhytismatales
Phomopsis velata	Phomopsis canker	*Abies, Larix, Pinus*	○	○	Diaporthaceae	Diaporthales
Cytospora pinastri	Cytospora canker	*Abies, Larix*	○	○	Valsaceae	Diaporthales
Valsa pini (doughtful)	(not proposed)	*Pinus*		○	Valsaceae	Diaporthales
Valsa salicina	Cytospora canker	*Pseudotsuga*		○	Valsaceae	Diaporthales
[b]*Cosmospora viridescens*	Nectria canker	*Alnus, Larix, Picea, Pinus*		○	Nectriaceae	Hypocreales
[b]*Thelonectria pinea*	(not proposed)	*Pinus*	○		Nectriaceae	Hypocreales
Mycodiella laricis-leptolepidis	needle cast	*Larix*		○	Mycodiella	Xylariales
[b]*Thyronectria balsamea*	(not proposed)	*Larix, Pinus*		○	Thyridiaceae	(Sordariomycetes)
[b]*Thyronectria cucurbitula*	Ophionectria canker	*Abies, Larix, Pinus*	○	○	Thyridiaceae	(Sordariomycetes)
Helicobasidium mompa	violet root rot	*Pinus*		○	Helicobasidiaceae	Helicobasidiales
Chrysomyxa abietis	needle cushion rust	*Picea*		○	Coleosporiaceae	Pucciniales
Chrysomyxa deformans	(not proposed)	*Picea*	○	○	Coleosporiaceae	Pucciniales
Coleosporium campanulae	needle rust	*Pinus*		○	Coleosporiaceae	Pucciniales
Coleosporium cimicifugatum	leaf rust	*Pinus*		○	Coleosporiaceae	Pucciniales

Scientific name	Disease name	Host			Family	Order
		Genus	Native	Introduced		
Coleosporium eupatorii	needle rust	*Pinus*		○	Coleosporiaceae	Pucciniales
[a]*Coleosporium neocacaliae*	needle rust	*Pinus*		○	Coleosporiaceae	Pucciniales
Coleosporium phellodendri	needle rust / rust	*Pinus, Phellodendron*		○	Coleosporiaceae	Pucciniales
[a]*Coleosporium yamabense*	needle rust	*Pinus*		○	Coleosporiaceae	Pucciniales
Cronartium orientale	gall rust (Asian pine-oak gall rust)	*Pinus*		○	Cronartiaceae	Pucciniales
Cronartium ribicola	blister rust	*Pinus*	○		Cronartiaceae	Pucciniales
[a]*Cronartium yamabense*	blister rust	*Pinus*	○		Cronartiaceae	Pucciniales
Melampsora larici-populina	needle rust	*Larix*		○	Melampsoraceae	Pucciniales
Melampsorella caryophyllacearum	witches' broom	*Abies*	○		Pucciniastraceae	Pucciniales
Melampsoridium hiratsukanum	rust	*Alnus*	○	○	Pucciniastraceae	Pucciniales
Pucciniastrum areolatum	cone rust	*Picea, Prunus*	○	○	Pucciniastraceae	Pucciniales
Pucciniastrum tiliae	rust / needle rust	*Abies, Tilia*	○	○	Pucciniastraceae	Pucciniales
Uredinopsis kameiana	needle rust	*Abies*	○		Pucciniastraceae	Pucciniales
Septobasidium kameii	felt disease	*Picea*	○	○	Septobasidiaceae	Septobasidiales
Armillaria mellea	Armillaria root rot	*Abies, Betula, Larix, Picea, Pinus*	○	○	Physalacriaceae	Agaricales

[a]New species described by Dr. Saho and Dr. Takahashi
[b]Species first reported in Japan by Dr. Saho and Dr. Takahashi

requiring further taxonomical reexamination. Moreover, the phylogenetic relationships between Japanese populations and European or American counterparts require further examination. Specimens collected by Dr. Saho and Dr. Takahashi through their surveys are stored in the herbarium of UTHF, which is entitled "TUFH".

Among these diseases, Takahashi (1979) listed the following five diseases as important ones that would cause serious forest damage at plantations: (1) Phacidium snow blight of *Picea jezoensis* and *Abies sachalinensis* caused by *Gremmenia abietis*, (2) Scleroderris canker of *A. sachalinensis* and *Pinus strobus* caused by *Gremmeniella abietina*, (3) Lachnellula canker of *P. strobus* caused by *Dasyscyphus pini*, (4) Septoria needle blight of *P. strobus* caused by *Septoria pini-pumilae*, and (5) Encoeliopsis canker of *Larix* spp., except for *L. kaempferi*, *L. gmelinii* var. *japonica*, and their hybrid, caused by *Gremmeniella laricina*. Takahashi (1979) also pointed out the following characteristics concerning the occurrence of conifer diseases:

- Disease occurrence is quite rare in natural forests, especially on seedlings and saplings.
- Most plantation diseases tend to occur frequently and severely at higher altitudes, while rust disease has the opposite tendency.
- The occurrence of the major diseases mentioned above is closely related to winter snow cover, at least except for Septoria needle blight.
- The disease risk for plantations is highest for *P. strobus*, high for *A. sachalinensis*, low for *Larix* species, and lowest for *P. glehnii*, especially at higher altitude.
- Many fungal species are observed both in plantations and natural forests, and it is suggested that pathogenic fungi inhabiting natural forests spread and cause outbreaks in nearby plantations; in other words, there is a possibility that disease in plantations originates from natural forests.
- Many fungal species are also distributed in European and North American boreal forests.

The plantation in Hokkaido district historically started with *L. kaempferi*, which led to needle cast (since 1935), Armillaria root rot (since 1954), and more severely, shoot blight (since 1960). Subsequently, the target species for the plantation shifted to *A. sachalinensis*, which faced issues with Scleroderris canker (since 1970), and later to *P. glehnii*. This shift in species choice has certainly been influenced by factors such as wood quality, but forest damages have also played a role in this shift. Artificial plantations of *P. jezoensis* until the 1960s were avoided because of late frost injury and damage caused by the gall-forming aphid *Adelges japonicus*. For the current plantation of hybrid larch in the UTHF, care should be taken against Armillaria root rot and needle cast, because it is relatively susceptible to these diseases due to the high susceptibility of its maternal *L. gmelinii* var. *japonica*.

14.2.3 Fungal Diseases That Occur on Picea jezoensis *in Relation to Its Growth Stage and Natural Regeneration*

Parasitic fungi on *P. jezoensis* in the natural forests were studied by Takahashi (1991) from the viewpoint of its life history, especially its natural regeneration. He identified several important diseases affecting different growth stages: cone rust (*Pucciniastrum areolatum*) for cones and seeds that inhibit seed maturation; Racodium snow blight (*Racodium therryanum*) and damping-off (*Fusarium* spp.) for seedlings; Racodium and Phacidium snow blights for saplings; Atropellis canker (*Discocainia treleasei*) for immature young trees; and wood decay (*Phellinus pini, Phaeolus schweinitzii, Laetiporus montanus, Heterobasidion parviporum, Fomitopsis pinicola*) for mature trees.

Takahashi (1991) also demonstrated that saplings on fallen logs were scarcely affected by snow blights, and that fallen logs provided safe sites for the natural regeneration of *P. jezoensis*. Wood-decaying fungi contributed to the formation of the "nursery-log regeneration system" by providing fallen logs on the forest floor (refer also to Chap. 12). Furthermore, he demonstrated that symbiotic ectomycorrhizal fungi (*Cenococcum graniforme* and *Thelephora terrestris*) played an important role in the growth of seedlings into saplings at nutrient-poor sites, including fallen logs. Snow molds that cause snow blight and wood-decaying fungi are described in the following sections.

14.2.4 Parasitic Wood-Decaying Fungi

In natural forests, particularly old growth forests, wood decay of living trees is the major forms of damage caused by fungi. Old-aged trees have been exposed to invasion by wood-rotting fungi through wounds and dead branches over extended periods, giving these fungi a significant time to decompose wood and spread to the tree body. Consequently, wood decay damage tends to be more pronounced in natural forests than plantations. Wood-rotting fungi that occur on living trees, including on the dead parts of living trees, in Hokkaido are listed in Table 14.2. This list is based on data summarized from Igarashi (2006), Takahashi (2007), Imazeki and Aoshima (1955), Kishi (1998), the Phytopathological Society of Japan (2015), and Kobayashi (2007). The scientific names conformed to the GBIF database (https://www.gbif.org/species/). However, further taxonomical and phylogenetic reexamination may be required for some species. For intance, *Heterobasidion annosum* from Asia was recently identified as *H. parviporum*, and *H. insulare* in Japan was reclassified as a new species, *H. orientale* (Tokuda et al. 2009). Additionally, *Laetiporus sulphureus* s. lat. found in Japan was reexamined, and Japanese *Laetiporus* species in cool temperate and boreal forests were identified as *L. cremeiporus* on broad-leaved trees and *L. montanus* on conifers (Ota et al. 2009). There is also a view that *Phellinus*

Table 14.2 Parasitic wood-decaying fungi recorded in Hokkaido district

Japanese name	Scientific name	Host	Part of rot		Type of rot	Family	Order
タモギタケ	*Pleurotus citrinopileatus*	*Acer, Fraxinus, Quercus, Ulmus*	–	–	White	Pleurotaceae	Agaricales
ニオイアミタケ	*Gloeophyllum odoratum*	*Abies, Picea*	Butt	Heart	Brown	Gloeophyllaceae	Gloeophyllales
キカイガラタケ	*Gloeophyllum sepiarium*	*Abies, Picea*	Stem	Heart	Brown	Gloeophyllaceae	Gloeophyllales
キチリメンタケ	*Gloeophyllum trabeum*	*Abies, Picea*	Stem	Heart	Brown	Gloeophyllaceae	Gloeophyllales
マツノウロコタケ	*Veluticeps abietina*	*Abies*	Stem	Heart	Brown (cubica/ honeycombed)	Gloeophyllaceae	Gloeophyllales
エゾノサビイロアナタケ	*Coniferiporia weirii*	*Abies, Larix, Picea*	Butt	Heart	White	Hymenochaetaceae	Hymenochaetales
ダイダイタケ	*Hymenochaete xerantica*	*Quercus*	Stem	Sap	White (pocket)	Hymenochaetaceae	Hymenochaetales
ヤケコゲタケ	*Inonotus hispidus*	*Fraxinus, Quercus*	Stem	–	White (spongy)	Hymenochaetaceae	Hymenochaetales
カワウソタケ	*Inonotus mikadoi*	*Alnus*	Stem	Heart	White (pocket)	Hymenochaetaceae	Hymenochaetales
カバノアナタケ	*Inonotus obliquus*	*Betula*	Stem	Heart	White	Hymenochaetaceae	Hymenochaetales
エゾカワラタケ	*Onnia circinata*	*Abies, Picea*	Stem	Heart	White	Hymenochaetaceae	Hymenochaetales
アズマタケ	*Onnia orientalis*	*Pinus*	Butt	Heart	White	Hymenochaetaceae	Hymenochaetales

Japanese name	Scientific name	Host	Part of rot		Type of rot	Family	Order
ニセカイメンタケ	*Onnia tomentosa*	*Abies, Larix, Picea*	Butt	Heart	White (spot)	Hymenochaetaceae	Hymenochaetales
エゾノカワラタケ	*Onnia triquetra*	*Picea*	Butt	Heart	White	Hymenochaetaceae	Hymenochaetales
エゾヒヅメタケ	*Phellinopsis conchata*	*Cercidiphyllum, Populus, Ulmus*	Stem	Heart	White	Hymenochaetaceae	Hymenochaetales
カラマツカタワタケ	*Phellinus chrysoloma*	*Larix*	Stem	Heart	White (pocket)	Hymenochaetaceae	Hymenochaetales
モミサルノコシカケ	*Phellinus hartigii*	*Abies, Picea*	Stem	Sap	White (fibrous)	Hymenochaetaceae	Hymenochaetales
キコブタケ	*Phellinus igniarius*	*Acer, Alnus, Betula, Populus, Quercus, Tilia*	Stem	Heart	White (spongy)	Hymenochaetaceae	Hymenochaetales
マツノカタワタケ	*Phellinus pini*	*Picea*	Stem	Heart	White (pocket)	Hymenochaetaceae	Hymenochaetales
サクラサルノコシカケ	*Phellinus pomaceus*	*Cerasus*	Stem	–	Brown	Hymenochaetaceae	Hymenochaetales
エゾノハスグサレタケ	*Phellopilus nigrolimitatus*	*Picea*	–	–	White (spot)	Hymenochaetaceae	Hymenochaetales
マクラタケ	*Pseudoinonotus dryadeus*	*Quercus*	Stem and Butt	Heart	White	Hymenochaetaceae	Hymenochaetales
エゾキコブタケ	*Sanghuangporus baumii*	*Syringa*	Stem	Heart	White	Hymenochaetaceae	Hymenochaetales
シロサルノコシカケ	*Oxyporus populinus*	*Acer*	Stem	Heart	White (fibrous)	Schizoporaceae	Hymenochaetales

(continued)

Table 14.2 (continued)

Japanese name	Scientific name	Host	Part of rot		Type of rot	Family	Order
シハイタケ	*Trichaptum abietinum*	*Abies, Picea*	Stem	Sap	White	–	Hymenochaetales
ミヤマシロアミタケ	*Antrodia heteromorpha*	*Abies, Fraxinus, Larix, Picea*	Stem	Heart	Brown	Fomitopsidaceae	Polyporales
シミガタセンベイタケ	*Calcipostia guttulata*	*Abies, Picea*	Stem	Heart	Brown	Fomitopsidaceae	Polyporales
エゾタケ	*Climacocystis borealis*	*Abies, Picea*	Butt	Heart	White (cubical)	Fomitopsidaceae	Polyporales
ホウロクタケ	*Daedalea dickinsii*	*Quercus*	Butt	Heart	Brown (cubical)	Fomitopsidaceae	Polyporales
カンバタケ	*Fomitopsis betulina*	*Betula*	Stem	Sap	Brown (crack)	Fomitopsidaceae	Polyporales
クロサルノコシカケ	*Fomitopsis castanea*	*Quercus*	Stem	Heart	Brown (cubical)	Fomitopsidaceae	Polyporales
エブリコ	*Fomitopsis officinalis*	*Larix, Picea*	Stem	Heart	Brown (cubical)	Fomitopsidaceae	Polyporales
ツガサルノコシカケ	*Fomitopsis pinicola*	*Abies, Larix, Picea*	Stem	Heart	Brown (cubical)	Fomitopsidaceae	Polyporales
ヤニタケ	*Ischnoderma resinosum*	*Abies, Acer, Fraxinus, Picea, Populus, Quercus, Tilia*	Stem	Heart	White	Fomitopsidaceae	Polyporales
マスタケ	*Laetiporus cremeiporus*	*Acer, Quercus*	Stem	Heart	Brown (cubical)	Laetiporaceae	Polyporales
アイカワタケ	*Laetiporus montanus*	*Abies, Larix, Picea*	Stem	Heart	Brown	Laetiporaceae	Polyporales
カイメンタケ	*Phaeolus schweinitzii*	*Abies, Larix, Picea*	Butt	Heart	Brown (cubical)	Fomitopsidaceae	Polyporales
レンゲタケ	*Postia balsamea*	*Abies, Larix, Picea*	Butt	Heart	Brown (cubical)	Fomitopsidaceae	Polyporales
オシロイタケ	*Postia tephroleuca*	*Acer, Salix*	Stem	Heart	Brown	Fomitopsidaceae	Polyporales

Japanese name	Scientific name	Host	Part of rot		Type of rot	Family	Order
ケニクアミタケ	*Rhodofomes cajanderi*	*Abies, Picea*	Stem	Heart	Brown (cubical)	Fomitopsidaceae	Polyporales
バライロサルノコシカケ	*Rhodofomes roseus*	*Abies, Picea*	Stem	Heart	Brown	Fomitopsidaceae	Polyporales
コフキタケ	*Ganoderma applanatum*	*Acer, Abies, Betula, Fraxinus, Populus, Tilia*	Stem	Heart	White	Ganodermataceae	Polyporales
マンネンタケ	*Ganoderma lucidum*	*Alnus, Betula, Quercus*	Butt	Heart	White	Ganodermataceae	Polyporales
マゴジャクシ	*Ganoderma neojaponicum*	*Abies, Picea*	Butt	–	White	Ganodermataceae	Polyporales
ツガノマンネンタケ	*Ganoderma valesiacum*	*Abies, Larix, Picea*	Butt	Heart	White	Ganodermataceae	Polyporales
マイタケ	*Grifola frondosa*	*Juglans, Quercus*	Butt	Heart	White	Grifolaceae	Polyporales
ニレサルノコシカケ	*Rigidoporus ulmarius*	*Ulmus*	Stem	Heart	White	Meripilaceae	Polyporales
エゾハリタケ	*Climacodon septentrionalis*	*Acer, Quercus, Ulmus*	Stem	Heart	White (spongy)	Phanerochaetaceae	Polyporales
アミヒラタケ	*Cerioporus squamosus*	*Tilia, Ulmus*	Stem	Heart	White	Polyporaceae	Polyporales
チャカイガラタケ	*Daedaleopsis tricolor*	*Acer, Alnus, Betula, Cerasus, Quercus*	Stem	Sap	White	Polyporaceae	Polyporales
カンバタケモドキ	*Favolus pseudobetulinus*	*Salix*	Stem	–		Polyporaceae	Polyporales
ツリガネタケ	*Fomes fomentarius*	*Acer, Betula, Tilia, Ulmus*	Stem	Heart	White	Polyporaceae	Polyporales

(continued)

Table 14.2 (continued)

Japanese name	Scientific name	Host	Part of rot		Type of rot	Family	Order
オオカボチャタケ	*Hapalopilus croceus*	*Quercus*	Stem	Heart	Brown	Polyporaceae	Polyporales
ハチノスタケ	*Lentinus arcularius*	*Abies, Morus*	Stem	–	White	Polyporaceae	Polyporales
カイガラタケ	*Lenzites betulinus*	*Abies, Quercus*	Stem	Heart	White	Polyporaceae	Polyporales
ニオイカワキタケ	*Neofavolus suavissimus*	*Salix, Sorbus*	–	–		Polyporaceae	Polyporales
エゾナミハタケ	*Neolentinus kauffmanii*	*Picea*	–	Heart	Brown (pocket)	Polyporaceae	Polyporales
ベッコウタケ	*Perenniporia fraxinea*	*Acer, Cerasus, Fraxinus, Ulmus*	Butt	Heart	White	Polyporaceae	Polyporales
キンイロアナタケ	*Perenniporia subacida*	*Abies, Cerasus, Larix, Picea*	Butt	Heart	White (fibrous)	Polyporaceae	Polyporales
ヒイロタケ	*Trametes coccinea*	*Abies, Acer, Betula, Quercus, Tilia*	Stem	Sap	White	Polyporaceae	Polyporales
オオチリメンタケ	*Trametes gibbosa*	*Acer, Betula, Tilia, Ulmus*	Stem	Sap	White	Polyporaceae	Polyporales
アラゲカワラタケ	*Trametes hirsuta*	*Abies, Acer, Alnus, Picea*	Stem	Sap	White (spongy)	Polyporaceae	Polyporales
シロアミタケ	*Trametes suaveolens*	*Fraxinus, Populus, Salix*	–	Heart	White	Polyporaceae	Polyporales
ウサギタケ	*Trametes trogii*	*Acer, Betula, Populus, Salix*	–	Heart	White	Polyporaceae	Polyporales
カワラタケ	*Trametes versicolor*	*Abies, Acer, Betula, Tilia, Ulmus*	Stem	Sap	White (spongy)	Polyporaceae	Polyporales

Japanese name	Scientific name	Host	Part of rot		Type of rot	Family	Order
ハナビラタケ	*Sparassis crispa*	*Abies, Larix, Picea*	Butt	Heart	Brown (cubical)	Sparassidaceae	Polyporales
ミヤマトンビマイタケ	*Bondarzewia mesenterica*	*Abies*	Butt	Heart	White	Bondarzewiaceae	Russulales
マツノネクチタケ	*Heterobasidion parviporum*	*Abies, Larix, Picea*	Butt	Heart	White (fibrous)	Bondarzewiaceae	Russulales
レンガタケ	*Heterobasidion orientale*	*Abies, Larix, Picea*	Butt	Sap	White (spongy)	Bondarzewiaceae	Russulales
ヤマブシタケ	*Hericium erinaceus*	*Quercus*	Stem	–	White	Hericiaceae	Russulales
キウロコタケ	*Stereum hirsutum*	*Acer, Quercus*	Stem	Sap	White	Stereaceae	Russulales
チウロコタケモドキ	*Stereum sanguinolentum*	*Abies, Larix, Picea*	Stem	Heart	White (fibrous)	Stereaceae	Russulales
カタウロコタケ	*Xylobolus frustulatus*	*Quercus*	Stem	Sap	White (pocket)	Stereaceae	Russulales
オオミコブタケ	*Kretzschmaria deusta*	*Acer, Quercus*	Butt	Heart	White	Xylariaceae	Xylariales

pini found in Japan is *P. yamanoi*, and the taxonomical status of Japanese *P. chrysoloma* remains unresolved (Sawahata and Hayashi 2002).

In the UTHF, a survey of wood-rotting fungi was conducted during 1946–1951 by Chiba and Teramoto (1952). They listed the following species as important wood-rotting fungi on living trees: *Phellinus hartigii, Phaeolus schweinitzii, Stereum sanguinolentum* and *Veluticeps abietina* for *A. sachalinensis*; *Phellinus pini, Phaeolus schweinitzii, Phellopilus nigrolimitatus, Onnia tomentosa* and *Perenniporia subacida* for *P. jezoensis* and *P. glehnii*; *Inonotus obliquus, Phellinus igniarius* and *Fomitopsis betulina* for *Betula* spp.; *Hapalopilus croceus, Laetiporus cremeiporus, Pseudoinonotus dryadeus, Phellinus igniarius, Inonotus hispidus* and *Xylobolus frustulatus* for *Quercus crispula*; *Rigidoporus ulmarius* and *Pleurotus citrinopileatus* for *Ulmus davidiana* var. *japonica*; *Phellinopsis conchata* for *Cercidiphyllum japonicum*; *Ganoderma applanatum, Oxyporus populinus* and *Climacodon septentrionalis* for *Acer* spp.; *Trametes trogii* and *T. suaveolens* for *Salix* and *Populus* spp.; *Fomes fomentarius* and *Fomitopsis castanea* for various broad-leaved species. The specimens collected by Chiba and Teramoto (1952) are stored in the herbarium of the Forest Botany Division, the University Museum, the University of Tokyo (coded as TOFO). Some major species found in the UTHF are shown in Fig. 14.10.

A detailed survey using sampling plots was also conducted by Chiba and Teramoto (1952) in forest compartment No. 63 at the time (Tairazawa), where a natural forest dominated by *A. sachalinensis* mixed with *P. jezoensis*, *Acer pictum*, *Fraxinus mandshurica* and *Ulmus davidiana* var. *japonica* was clear cut. They

Fig. 14.10 Fruit bodies of some major parasitic wood-decaying fungi. (All photos by Daisuke Sakaue). *Phellinus hartigii* (**a**) and *Heterobasidion orientale* (**b**) on *Abies sachalinensis*, *Phellinus pini* (**c**), *Fomitopsis pinicola* (**d**), *Heterobasidion parviporum* (**e**), *Phaeolus schweinitzii* (**f**) and *Onnia tomentosa* (**g**) on *Picea jezoensis*, *Fomes fomentarius* on *Betula platyphylla* var. *japonica* (**h**) and *B. ermanii* (**i**), *Laetiporus cremeiporus* (**j**) on *Cercidiphyllum japonicum*

reported that wood decay reached 54.7% in tree number and 14.9% and 13.9% in volume of coniferous and broad-leaved trees, respectively. They also noted that decay was conspicuous in small-sized *A. sachalinensis*. Other sampling plot surveys were conducted by Sasaki and Yokota (1955, 1956) in *A. sachalinensis*-dominated natural forests of forest compartments Nos. 55 and 56 at the time (Kawamatsuzawa), No. 53 (Sasazawa), No. 47 (Sannoyama), and No. 50 (Iwanazawa). They reported that 43.5–79.8% of *A. sachalinensis* trees (5.8–7.1% in volume) decayed, and *Heterobasidion parviporum* and *Postia balsamea* were added as important parasitic wood-rotting fungi on *A. sachalinensis*. There is a unique type of *A. sachalinensis* whose bark is brownish, split small and tortoise shell-like, and sometimes accompanied by black gall, while the common bark of *A. sachalinensis* is grayish white and smooth. In this type of *A. sachalinensis* grown in forest compartments No. 16 (Mizunashizawa) and No. 7 (Shinkomatsuzawa), a larger number of trees decayed, and the proportion reached 82.0–91.0% (Yokota 1957).

The surveys mentioned above were all conducted at lower altitudes. At higher altitudes of forest compartment No. 14, the proportions of decayed trees were 31% and 51% for *P. jezoensis* and *A. sachalinensis*, respectively. Similar to Chiba and Teramoto (1952) and Sasaki and Yokota (1955, 1956), many decayed *A. sachalinensis* were observed even in small trees (Sakaue, unpublished data).

Thus, the wood decay of living trees is prominent in natural forests and should be considered for natural forest management. For example, the "pathological cutting age" for each tree species should be discussed. This refers to determining the optimal cutting age from the perspective of minimizing timber volume losses due to wood decay. In the practice of forest management, the skill for diagnosing wood decay inside trees is very important. However, this skill currently relies heavily on the practical knowledge of highly experienced foresters (technical staff). It is therefore essential to extract, validate, and share this empirical knowledge, as well as to develop and improve diagnostic technologies and tools.

14.2.5 Snow Molds and Attempts to Control Them

Fungi that grow and attack dormant plants at low temperatures under snow cover are called snow molds. Two types of snow molds that affect conifers are known in the world (Sinclair and Lyon 2005). One is the type that produces annual mycelia and causes "snow blight," and the major species are *Gremmenia infestans* in Europe, *Gremmenia abietis* in North America, and *Lophophacidium hyperboreum* in Eurasia and North America. Another type of fungi, *Herpotrichia pinetorum* and *Neopeckia coulteri*, cause "brown felt blight" with perennially growing brown mycelia. *G. infestans* and *G. abietis* were formerly known as members of the genus *Phacidium*, and *H. pinetorum* was formerly known as *H. juniperi*.

In Japan, the major snow molds are *Racodium therryanum* at lower altitudes, which causes Racodium snow blight (Fig. 14.11), and *Gremmenia abietis* at higher altitudes, which causes Phacidium snow blight. Two additional species, *Botrytis*

Fig. 14.11 Racodium snow blight on *Picea jezoensis*. (**a**: photo by Daisuke Sakaue, May 17, 2011), *Picea glehnii*. (**b**: photo by Daisuke Sakaue, May 17, 2011), *Abies sachalinensis*. (**c**: photo by Daisuke Sakaue, May 17, 2011), and *Cephalotaxus harringtonia* var. *nana*. (**d**: photo by Daisuke Sakaue, May 2, 2018)

cinerea (gray mold) and *Sclerotinia kitajimana* (screlotial disease), have also been reported as snow molds, but damage caused by these two species is rare in Hokkaido (Sato et al. 1959). *R. therryanum* is phylogenetically thought to be *H. pinetorum*, but its life cycle is greatly different from that of Eurasian and North American *H. pinetorum*, because the mycelia are annual and ascospores are not found (Iwakiri et al. 2021). On the other hand, *G. abietis* in Japan was recently described as a new species, *Chionobium takahashii*, which belongs to a new genus, neither *Gremmenia* nor *Phacidium* (Iwakiri et al. 2024). Thus, the taxonomical position and pathological characteristics of Japanese snow molds on woody plants is still controversial.

As mentioned in the above sections, snow molds are fatal to juvenile trees. Among them, *Racodium therryanum* is very important, because it kills seeds, seedlings, and saplings and disturbs natural regeneration, particularly that of *P. jezoensis*. On the other hand, the importance of *Gremmenia abietis* on the natural regeneration of conifers is not well understood. While *R. therryanum* has a wide host range from conifers to broad-leaved trees, the susceptibility of each species greatly differs between seeds and seedlings: *P. jezoensis* is more susceptible than *A. sachalinensis* in seedlings, but the opposite is true for seeds, and seeds of *P. glehnii* are much more susceptible than those of *A. sachalinensis* compared to seedlings (Cheng and Igarashi 1990). Therefore, *R. therryanum* inhibits the natural regeneration of not only *P. jezoensis* but also *A. sachalinensis* (Hayashi and Endo 1975) and possibly *P. glehnii* during the process of emergence by affecting their seeds.

To avoid damage caused by snow molds and facilitate natural regeneration, various attempts have been conducted in the UTHF. One approach is the artificial control of snow depth and snow-covering duration at the pre-emergence stage of seeds. Among several snow control treatments, field trials have demonstrated that the artificial formation of uneven ground surfaces enhances soil frost during winter in natural forests (Owari et al. 2014). An inoculation test of *A. sachalinensis* and *P. jezoensis* seeds with *R. therryanum* showed that the rates of both fungal infection and germinability loss decreased under such treatment, and additionally, the infection rate of *Betula maximowicziana* seeds decreased (Sakaue et al. unpublished data). An increase in soil frost may be effective in preventing the seed loss of diverse tree species caused by snow molds. Another approach is the installation of materials that assist natural regeneration on forest floors as seedbeds. This means that they are alternatives to nursery logs and stumps. However, effective materials and methods have not yet been developed. Although more germinated seedlings were obtained on some materials, such as decayed wood, lapilli, and by the treatment which shut fallen leaves out, most of them deceased rapidly after the first winter due to frost heaving that dug seedlings out and soil splashes that covered the seedlings, as well as due to snow molds (Takahashi 1991; Kamoda et al. 2010). Further trials that ensure seedling establishment are needed.

14.3 Forest Damage Caused by Mammals

In Hokkaido, including introduced species, there are 53 species of terrestrial mammals. Of these, only a few, such as deer, hare, and vole, are considered to cause damage to forestry. However, the damage to forestry caused by these mammals can be significant depending on the era, and considerable effort has been made to counteract this damage.

14.3.1 Rodent

In Hokkaido, forestry operations began around the end of the nineteenth century when the UTHF was established. When larch trees started to be planted, damage caused by the Ezo red-backed vole, *Craseomys rufocanus,* was observed from the early stages of these operations. In 1937, rodent damage affected 10,000 hectares of afforested land throughout Hokkaido (Ueda et al. 1966). This was the first time such damage occurred on a prefecture-wide scale, and the total cost of the damage reached 200,000 yen at that time. Subsequently, in 1942, 1943, 1951, 1959, and

1964, major damages occurred across Hokkaido, particularly in 1959, when the affected area extended to as much as 120,000 hectares (Ueda et al. 1966).

With the occurrence of damage caused by the vole, research on the mechanisms of occurrence and control measures began to be conducted in various parts of Hokkaido. The content ranged widely and included ecological studies, repellent research, chemical control, biological control, mechanical control, and forestry control. In the UTHF, various tests and studies have focused on trials of vole resistance among larch tree species and varieties. This was because one of the priority research projects was to cultivate superior reforestation lineages with high vole tolerance through interspecific hybridization of larch species (see Chap. 10).

Takahashi and Iwamoto (1963) established three plots for damage experiments at the UTHF in April 1962. In each block, they planted seedlings of coniferous tree species from both domestic and overseas areas, broadleaf tree seedlings, and seedlings of various *Larix* species. This study aimed to investigate the damage rate under conditions where there was a relatively high population density of the vole. In the conifer planting blocks, Manchurian black pine, *Pinus tabuliformis* (98.2%), unknown pine species (Parashiyāna-matsu in Japanese) (95.6%), mountain pine, *Pinus mugo* (92.8%), *L. kaempferi* (91.8%), and *Pinus sylvestris* (90.8%) suffered severe damage. Additionally, *Pinus strobus* (50.5%), black spruce, *Picea mariana* (36.1%), white spruce, *Picea glauca* (19.9%), and Olgan larch, *Larix olgensis* var. *koreana* (17.2%) experienced moderate damage. On the other hand, *P. glehnii*, *A. sachalinensis*, and *L. gmelinii* var. *japonica* suffered minimal damage. In the broadleaf tree planting block, *B. maximowicziana* (47.6%), Japanese poplar, *Populus suaveolens* (28.6%), alder, *Alnus inokumae* (19.7%) and aspen, *Populus tremula* var. *sieboldii* (17.5%) suffered moderate damage. Conversely, Japanese white birch, *Betula platyphylla* var. *japonica* (5.3%), Manchurian alder, *Alnus hirsuta* (4.7%), *Fraxinus mandshurica* (2.9%) and Japanese alder, *Alnus japonica* (2.9%) suffered low damage. In the planting block of the genus Larix, *L. kaempferi* suffered severe damage at 90.9%, followed by Siberian larch, *L. sibirica* at 64.2%. The F_1 hybrids of *L. olgensis* var. *koreana* and *L. kaempferi* exhibited moderate damage (47.8%), while *L. olgensis* var. *koreana* exhibited 19.2% damage. On the other hand, the F_1 hybrid of *L. gmelinii* var. *japonica* and *L. kaempferi* suffered low damage at 3.6%, while *L. kaempferi* suffered minimal damage at 0.6%.

Takahashi and Nishiguchi (1966a) evaluated the resistance of various tree species to vole by examining the feeding preferences of the Ezo red-backed vole on seedlings of coniferous trees (larch, pine, fir, and spruce) from 1958 to 1965. In feeding experiments within the genus *Larix*, the results revealed that *L. kaempferi*, European larch, *L. decidua*, and western larch, *L. occidentalis* were preferred by the Ezo red-backed vole, indicating low resistance to voles. On the other hand, *L. olgensis* var. *koreana* and *L. laricina* demonstrated high resistance to voles. Additionally, *L. gmelinii* var. *japonica* from the Kuril Islands and Sakhalin exhibited extremely high resistance to voles. According to the results of experiments within the genus *Pinus*, Japanese Red Pine, *Pinus densiflora*, Japanese Black Pine, *P. thunbergii*, *P. sylvestris*, Red Pine, *P. resinosa*, *P. banksiana*, Pitch Pine, *P. rigida*, Loblolly Pine, *P. taeda*, and Himalayan Pine, *P. wallichiana* are considered to have low resistance to vole, which is thought to be equal to or less than that of *L. kaempferi*. On the other hand, the resistance of *P. strobus* was considered to be greater than that of *L. kaempferi*. The resistance of Korean pine, *P. koraiensis* and Japanese white pine, *P. parviflora* var. *parviflora* was intermediate between that of *L. olgensis* var. *koreana* and *L. gmelinii* var. *japonica*. In experiments with seven species of the genus *Picea*, Serbian spruce, *P. omorika* had the lowest resistance to vole, while *P. jezoensis* exhibited the highest resistance. However, none of the species could be considered to have exceptionally high resistance to vole. In experiments with six species within the genus *Abies* (and Douglas Fir, *Pseudotsuga menziesii*), it was found that concolor fir, *A. concolor* had very low resistance to vole. The resistance of balsam fir, *A. balsamea*, grand fir, *A. grandis*, silver fir, *A. alba*, Nikko fir, *A. homolepis*, and *P. menziesii* was comparable to that of *P. strobus*. On the other hand, the resistance of *A. sachalinensis* was high. Based on these results, tree species from Hokkaido, Sakhalin, the Kuril Islands, and Korea, such as *L. olgensis* var. *koreana*, *L. gmelinii* var. *japonica*, *P. koraiensis*, *P. parviflora* var. *parviflora*, *P. jezoensis*, *P. glehnii*, and *A. sachalinensis*, exhibit high resistance to vole, whereas most foreign tree species and those native to Honshu, such as *P. densiflora*, *P. thunbergii*, *L. kaempferi*, and *A. homolepis*, do not show high resistance.

Takahashi and Nishiguchi (1966b) conducted experiments on resistance to vole for hybrid species between *L. olgensis* var. *koreana* and *L. kaempferi*. Their results showed that the resistance to vole of F_1 hybrids is intermediate between that of both parents, with a bias toward the maternal side. In addition, it has been suggested that F_1 hybrids with *L. gmelinii* var. *japonica* as the mother exhibit high resistance to vole, indicating their potential as promising tree species for afforestation (refer to Chap. 10). Ogasawara et al. (1973, 1974) evaluated resistance to voles in hybrids of *L. gmelinii* var. *japonica* and *L. kaempferi,* including F_1 and F_2 generations, as well as in backcrossed progenies. They found that the vole resistance in the F_1 generation was closer to that of the stronger parent species. The F_2 generation was slightly inferior to the F_1 generation, but the backcrossed lineages leaned toward the side of the backcrossed parent species. Hybrids of Kuril's *L. gmelinii* var. *japonica* and *L. kaempferi* demonstrated significantly greater vole resistance compared to those between Sakhalin's *L. gmelinii* var. *japonica* and *L. kaempferi*.

14.3.2 Sika Deer

As the planted trees grew, vole damage decreased, and research on vole resistance in Hokkaido became less common. However, in the 1990s, there was a rapid increase in the sika deer, *Cervus nippon,* population, leading to evident feeding damage by them (Fig. 14.12). Takahashi et al. (1997) reported on bark stripping by sika deer in 1996 within the Iwanazawa LTER plot (refer to also Chap. 5). According to their report, out of 43 tree species growing in the area, 24 species suffered some form of feeding damage. The main species that suffered more than 50% feeding damage were *Taxus cuspidata* (52.6%) and *Ulmus laciniata* (51.5%). Over 10% of the other

Fig. 14.12 Nikko fir, *Abies homolepis,* trees damaged by deer at an experimental site in the arboretum of the UTHF. (Photo by Dai Fukui, March 2022)

species, such as *Ulmus davidiana* var. *japonica* (28.9%), *Morus australis* (15.8%), *Hydrangea paniculata* (14.0%), *Cornus controversa* var. *controversa* (12.9%), *Padus ssiori* (12.6%), *A. sachalinensis* (10.6%), and *Salix* spp. (11.3%), were damaged. In particular, for *U. laciniata* and *U. davidiana* var. *japonica*, there was a high percentage of trees with complete bark stripping around their trunks, resulting in 70–80% mortality among the damaged trees. A damage survey in the Iwanazawa LTER plot continued even after, and as of 2009, the damage rates were 85.3% for *T. cuspidata*, 83.6% for *U. laciniata*, 65.8% for *H. paniculata*, and 41.5% for *U. davidiana* var. *japonica*. Furthermore, the mortality rates reached 92.6% for *U. laciniata* and 96.2% for *H. paniculata*. Within the 18.75-hectare plot, the number of surviving trees with a diameter greater than 5 cm decreased from 10,437 at the time of establishment (1994) to 7184 over 25 years—a reduction of 69% (Ogawa et al. 2022). Additionally, in *U. davidiana* var. *japonica* and *U. laciniata*, which have high browsing rates, smaller-diameter individuals were more susceptible to damage, consistent with the results of Suzuki et al. (2011). In other words, the forest composition and size structure in the Iwanazawa LTER plot dramatically changed due to the impact of deer. However, damage rates from 2009 to 2019 have been decreasing: 22% for *T. cuspidata*, 9% for *U. laciniata*, and 10% for *U. davidiana* var. *japonica* (Ogawa et al. 2022). Attention will be focused on whether the populations of species that have declined significantly will recover.

According to Kaji (1997), a population survey conducted by Hokkaido in 1996 using helicopters revealed relatively high population densities in different regions. In the Yamabe region: 7.10 individuals/km^2. In the Nunobe region: 5.06 individuals/km^2. In three sections of the middle reaches of the Nishitappu River: 10.96, 12.54, and 5.47 individuals/km^2, respectively.

Subsequently, research on the feeding preferences of sika deer and studies related to bark-stripping risk factors were conducted at the UTHF. Yasui et al. (2002) conducted experiments on the bark preference of 12 tree species and revealed that Sika deer have a high preference for *U. laciniata*, *T. cuspidata*, *F. mandshurica*, and *U. davidiana* var. *japonica* but a low preference for *B. platyphylla* var. *japonica*, *Betula maximowicziana*, *Maackia amurensis*, and *Acer pictum*. Suzuki et al. (2011) investigated the impact of individual tree attributes and surrounding environmental conditions on the risk of bark stripping per individual. They found that, in general, smaller-diameter trees have a greater probability of bark stripping. Furthermore, convex landforms and a high proportion of supplementary plantations near habitats positively influence the probability of bark stripping. These environmental conditions align with those utilized by deer during snowmelt periods, suggesting that the risk of bark stripping in the UTHF increases in locations frequently used by deer during these periods.

In the UTHF, with the increase in the number of sika deer population, light censuses have been conducted for three consecutive nights every November since 2007. These censuses are carried out both within the UTHF and in agricultural lands adjacent to the UTHF. The results show that in the forest, the average number of deer sightings has remained stable at less than 20 individuals per km. However, on farmlands, the numbers fluctuate significantly and irregularly, ranging from as high as 90 individuals per km in some years to less than 10 individuals per km in others (The University of Tokyo Hokkaido Forest 2017). Considering that an average sighting number of less than 20 individuals per 10 km is categorized as a low-density area, and 20–100 individuals per 10 km are categorized as a medium-density area, it can be inferred that while forests maintain a stable low density, farmlands experience substantial fluctuations between low and medium density. However, even within the UTHF itself, there appear to be spatiotemporal fluctuations in density. For instance, around the Iwanazawa LTER plots mentioned earlier, it is estimated that during the 1990s–2000s, the density was severe enough to cause significant forest damage. Such spatiotemporal fluctuations in density will continue to occur in the future, necessitating ongoing monitoring and appropriate measures based on the results.

References

Arisawa H, Shibano S, Iguchi K, Nihara K, Ohta S, Furuta K (1986) An outbreak of *Dendrolimus spectabilis* on *Pinus strobus* in Hokkaido. Proc Annu Meet Jpn For Soc 97:493–494. (in Japanese)

Arisawa H, Iguchi K, Shibano S (1987) Mortality of overwintered *Dendrolimus spectabilis* larvae – Post outbreak year and two years after. Trans Meet Hokkaido Branch Jpn For Soc 36:131–133. (in Japanese)
Cheng D, Igarashi T (1990) Comparison of susceptibility to *Racodium therryanum* infection of seeds and seedlings of four coniferous species. Res Bull Coll Exp For Hokkaido Univ 47:125–136. (in Japanese with English summary)
Chiba O, Teramoto T (1952) Wood-rotting fungi collected in the Tokyo University Forest in Hokkaido. Bull Tokyo Univ For 43:19–37. (in Japanese with English summary)
Elton CS (1958) The ecology of invasions by animals and plants. Methuen, London
Forest Research Department, Hokkaido Research Organization (FRD HRO) (1995) *Operophtera brumata.* https://www.hro.or.jp/list/forest/research/fri/zukan/konchu/00data/cho-ga/shakuga/namisuji/note.html (Last visited on AUG01, 2023)
Forest Research Department, Hokkaido Research Organization (FRD HRO) (2010) *Pristiphora erichsoni.* https://www.hro.or.jp/list/forest/research/fri/zukan/konchu/00data/hati/habati/kara-hara/note.html. Last visited on 1 Aug 2023
GBIF. The Global Biodiversity Information Facility database. https://www.gbif.org/species/
Genebank Project, NARO. Japan Common name database of plant diseases in Japan. https://www.gene.affrc.go.jp/databases-micro_pl_diseases.php. Updated 2023/3/15, viewed 2023/7/7
Hara H, Hayashi N (2002) The degree of infestation of the bark beetle, *Ips typographus japonicus* NIIJIMA, after thinning in the young man-made forests of Sakhalin spruce, *Picea glehnii* MASTERS. Bull Hokkaido For Res Inst 39:69–74. (in Japanese with English summary)
Hayashi K, Endo K (1975) Effect of soil fungi and drought on the germination of Todo-fir seed. Bull Gov For Exp Sta 274:1–22. (in Japanese with English summary)
Igarashi T (2006) Fungi of Hokkaido. (in Japanese)
Iguchi K (2023) Observation of insect damages in the University of Tokyo Hokkaido Forest – 43 years since 1980. For Pests (Jpn) 72:42–48
Iguchi K, Ozaki K (2006) Three-year periodic outbreak of *Parana kamijoi* Togashi caused by prolonged diapause. Trans Meet Hokkaido Branch Jpn For Soc 54:121–123. (in Japanese)
Iguchi K, Yamamoto H, Furuta K (1993a) The spread of attack of Yezo-spruce by spruce beetles in natural forests after selection cutting. Bull Univ Tokyo For 90:1–5. (in Japanese)
Iguchi K, Yamamoto H, Furuta K (1993b) The spread of attack of Yezo-spruce by spruce beetles in natural forests after selection cutting. Bull Tokyo Univ For 90:1–15. (in Japanese with English summary)
Iguchi K, Takahashi Y, Inukai M, Kaji M (1998) Foliage damage of two species of *Tilia* (*T. japonicus* and *T. maximowicziana*) by *Parna* spp. Trans Meet Hokkaido Branch Jpn For Soc 46:117–120. (in Japanese)
Iguchi K, Yamamoto H, Furuta K (1999) Age of yezo-spruce killed by beetles and their growth before and after selection cuttings. Bull Univ Tokyo For 101:1–9. (in Japanese)
Iguchi K, Ozaki K, Isono M (2000) Life history of *Parana kamijoi* Togashi in central Hokkaido. Trans Meet Hokkaido Branch Jpn For Soc 49:96–98. (in Japanese)
Imazeki R, Aoshima K (1955) Fungus decay. In: A memoir of the scientific investigations of the primeval forests in the headwaters of the river Ishikari, Hokkaido, Japan, pp 239–271. (in Japanese with English summary)
Iwakiri A, Matsushita N, Fukuda K (2021) Snow mold fungus *Racodium therryanum* is phylogenetically *Herpotrichia juniperi*. Mycoscience 62:406–409
Iwakiri A, Hirooka Y, Matsushita N, Fukuda K (2024) *Chionobium takahashii*, gen. et sp. nov., associated with snow blight of conifers in Japan. Mycologia 116:299–308
Kaji M (1997) Sika ni yoru shinrin higai – Tokyo daigaku ensyurin no jirei (Forest damage by sika deer: a case study in the University of Tokyo Forests). Hoppou Ringyo 49:169–172
Kamata N (2002) Outbreaks of forest defoliating insects in Japan, 1950-2000. Bull Entomol Res 92:109–118
Kamata K, Iguchi K, Sanguansub S, Maneerat T (2014) Infestation by insect borers on *Betula maximowicziana* following successive years of severe defoliation by *Caligula japonica*

(Lepidoptera: Saturniidae) with special reference to *Xyleborus seriatus* Blandford (Coleoptera: Curculionidae: Scolytinae). For Pests (Jpn) 72:42–48

Kamoda S, Sakaue D, Goto S, Inoue H, Kusumoto D, Ogawa H, Inukai H, Kamata N, Yamada T, Kaji M (2010) Some artificial treatments and seedbed materials affect the survival and growth of Yezo spruce seedlings. Trans Meet Hokkaido Br Jpn For Soc 58:135–137. (in Japanese)

Kishi K (ed) (1998) Plant diseases in Japan. (in Japanese)

Kobayashi T (2007) Index of fungi inhabiting woody plants in Japan – host, distribution and literature.

Matsui M, Okawa A, Nakagawa Y (2017) Ezoshika niyoru ettouchi deno jumoku higai -Iwanazawa daimenseki chouki seitaikei plot niokeru higaijirei- [Tree damage caused by sika deer in wintering grounds - A case study of damage in the Iwanazawa LTER plot]. FY2017 Report on Research and Training Conference for Technical Staff of the University of Tokyo Forests pp 15–20. (In Japanese)

Nakayama M, Furuta K, Takahashi I, Sato Y, Iguchi K (1991) Death of Ezo-spruce around log depots after partial cuttings and the dynamics of adult spruce beetles. Bull Tokyo Univ For 84:39–52. (in Japanese with English summary)

Nitto M, Akui Y (1957) The spruce web-spinning sawfly (*Cephalcia isshikii* TAKEUCHI) in the University Forest in Hokkaido. Misc Inf Univ Tokyo For 12:37–67

Ogasawara S, Takahashi N, Kurahashi A, Hamaya T (1973) Vole-resitance of hybrids between Kurile and Japanese larch. Trans Meet Hokkaido Branch Jpn For Soc 22:115–121. (In Japanese with English summary)

Ogasawara S, Kurahashi A, Hamaya T (1974) Vole resistance of hybrids between Kurile and Japanese larch (continued). Trans Meet Hokkaido Branch Jpn For Soc 23:47–49. (In Japanese)

Ogawa H, Ohkawa A, Iguchi K, Suzuki SN (2022) 25-year dynamics of a riparian forest in an 18 ha Long-Term Ecological Research plot. Boreal For Res 70:49–52. (In Japanese)

Ota Y, Hattori T, Banik MT, Hagedorn G, Sotome K, Tokuda S, Abe Y (2009) The genus *Laetiporus* (Basidiomycota, Polyporales) in East Asia. Mycol Res 113:1283–1300

Owari T, Sakaue D, Shibano H, Miyamoto T (2014) Effects of snow control treatments on snow cover and soil frost in natural forest: Towards the development of snow control techniques that facilitate natural tree regeneration. Boreal For Res 62:17–18. (in Japanese)

Phytopathological Society of Japan (2015) Common names of plant diseases in Japan.

Pinkantayong P, Kubo M, Terada T, Fujii M, Kamoda S, Muramoto K, Kamata N (2015) The effect of successive years of defoliation by the larch sawfly (*Pristiphora erichsonii* (Hartig)) on the foliage properties of the Japanese larch (*Larix kaempferi* (Lamb.) Carr.), with particular reference to the CN balance hypothesis. J For Res 20:104–113

Root RB (1973) Organization of a plant-arthropod association in simple and diverse habitats: the fauna of Collards (*Brassica Oleracea*). Ecol Monogr 43:95–124

Saho H, Takahashi I (1973) A check list and host index of fungi on forest trees collected at the Tokyo University Forest in Hokkaido, Japan (1960-1972). Jubilee Volume in Commemoration of the Seventieth Birthday of Dr. Naohide HIRATSUKA (Rept). Tottori Mycol Inst (Jpn) 10:703–714

Sasaki T, Yokota S (1955) Wood decay of *Abies sachalinensis* forest in Tokyo University Forest in Hokkaido II. Bull Tokyo Univ For 52:75–87. (in Japanese with English summary)

Sasaki T, Yokota S (1956) Wood decay of *Abies sachalinensis* forest in Tokyo University Forest in Hokkaido. Misc Inform Tokyo Univ For 10:15–21. (in Japanese with English summary)

Sato K, Shoji T, Ota N (1959) Studies on the snow molding of coniferous seedlings –I Gray mold and sclerotial disease. Bull Gov For Exp Sta 110:1–153

Sawahata T, Hayashi Y (2002) Pairing tests and morphological studies of *Phellinus pini*-complex in Japan. Trans Mycol Soc Japan 43:51–59. (in Japanese with English summary)

Shibata M, Iguchi K (1981) *Kalopanax septemlobus* defoliation by a Hesperiid butterfly *Bibasis aquilina*. Hoppo-Ringyo (Northern For) 33:93–95

Sinclair WA, Lyon HH (2005) Diseases of trees and shrubs, 2nd edn.

Suzuki M, Fujiwara A, Kamoda S, Maehara T, Saito H, Matsui M, Iguchi K, Kaji M, Kamata N (2011) Factors affecting the probability of bark-stripping by deer in a natural forest; an analysis based on tree characteristics, deer habitat utilization and environmental conditions. J Jpn For Soc 93:213–219

Takahashi I (1979) Studies on mycoflora and diseases of coniferous trees at the central part of Hokkaido, Japan – special reference to Ascomycetes, Fungi Imperfecti and Uredinales. Bull Tokyo Univ For 69:1–143. (in Japanese with English summary)

Takahashi I (1991) Studies on the mycofloral succession on Yezo spruce (*Picea jezoensis* CARR.) in various stages of growth, with special reference to the role of fungi in the natural regeneration of the host. Bull Tokyo Univ For 86:201–273. (in Japanese with English summary)

Takahashi I (2007) Fungi of Hokkaido, enlarged new ed. (in Japanese)

Takahashi N, Iwamoto M (1963) Shuyou zourin jushu no yaso, yato no yagai shokugai jikken [Field experiments on feeding damage to major afforestation tree species by wild vole and hare]. Trans Meet Hokkaido Branch Jpn For Soc 12:102–109. (In Japanese)

Takahashi N, Nishiguchi C (1966a) Studies on the resistance of forest trees to the red-backed vole, *Clethrionomys rufocanus bedfordiae* (Thomas). (I) Relative feeding preference of the vole for seedlings of conifers under laboratory conditions. Bulle Tokyo Univ For 62:153–172. (In Japanese with English summary)

Takahashi N, Nishiguchi C (1966b) Studies on the resistance of forest trees to the red-backed vole, *Clethrionomys rufocanus bedfordiae* (Thomas). (2) Relative feeding preference of the vole for seedlings of larch F_1 hybrids under laboratory conditions. Bull Tokyo Univ For 62:173–188. (In Japanese with English summary)

Takahashi Y, Inukai M, Iguchi K, Takahashi I (1997) A case of forest damage caused by sika deer (*Cervus nippon yesoensis* Heude) in Iwanazawa experimental plot of the Tokyo University Forest in Hokkaido. Trans Meet Hokkaido Branch Jpn For Soc 45:84–87. (In Japanese)

The University of Tokyo Hokkaido Forests (2017) Ezo-shika light census 10 nen no ayumi [10-year history of the sika deer light census]. Univ Tokyo For News 77:3

Tokuda S, Onodera K (2018) Bark beetle infestation and Armillaria root rot occurred in a larch stand in eastern Hokkaido. Jpn Boreal For Res 66:15–18. (in Japanese)

Tokuda S, Hattori T, Dai Y-C, Ota Y, Buchanan PK (2009) Three species of *Heterobasidion* (Basidiomycota, Hericiales), *H. parviporum, H. orientale* sp. nov. and *H. ecrustosum* sp. nov. from East Asia. Mycoscience 50:190–202. https://doi.org/10.1007/s10267-008-0476-7

Ueda M, Higuchi S, Igarashi B, Maeda M, Kuwahata T, Ota K, Abe H, Fujimaki Y, Fujikura J, Takayasu T (1966) Historical review of studies on the Bedford's red-backed vole, *Clethrionomys rufocanus bedfordiae* (Thomas). Bull Gov For Exp Sta 191:1–100. (In Japanese)

Watanabe S, Shibata S, Kawahara S, Shibano S, Kurahashi A, Satoo Y, Anazawa C, Takada N, Takahashi Y (1990) A memoir on the actual situation of the forest wind-damaged by the typhoon No. 15 in 1981 in the Tokyo University Forest in Hokkaido. Misc Inf Univ Tokyo For 27:79–221

Yamaoka Y, Takahashi I, Iguchi K (2000) Virulence of ohiostomatoid fungi associated with the spruce bark beetle *Ips typographus f. japonicas* in Yezo spruce. J For Res 5:87–94

Yasui Y, Orihashi K, Kojima Y, Terazawa M (2002) Bark selection of free ranging sika deer (*Cervus nippon yesoensis* Heude): cafeteria tests during snow season. Trans Meet Hokkaido Branch Jpn For Soc 50:79–81. (In Japanese)

Yokota S (1957) Wood decay of Abies sachalinensis forest in the Tokyo University Forest, Hokkaido III. On-wood decay of Oni-todo (*Abies sachalinensis* MAST. of bad-looking bark). Bull Tokyo Univ For 53:139–148. (in Japanese with English summary)

Chapter 15
Biodiversity Conservation

Dai Fukui and Satoshi N. Suzuki

Abstract The University of Tokyo Hokkaido Forest (UTHF) manages extensive areas of mature natural forests using the Stand-based Silvicultural Management System, which provides a favorable environment for forest organisms. In addition to forest reserves, where no silvicultural management or logging is conducted, there are other areas where logging is also not carried out for various reasons. A wildlife protection area including special protection zone is also designated within the UTHF. These measures contribute to the preservation of diverse species' habitats. However, the recent increase in deer populations poses a risk to biological conservation due to excessive feeding damage. Climate change also presents a potential threat, particularly to alpine species. Furthermore, the rising numbers of nonnative species, such as the raccoon dog, may cause additional risks.

Keywords Forest reserve · Restricted forest · Protection area · Endangered species · Nonnative species

The University of Tokyo Hokkaido Forest (UTHF) has been managed for more than 60 years based on the Stand-based Silvicultural Management System (SSMS) principles (refer to Chap. 7) of "facilitating the transition to and sustaining a pre-climax phase for each stand" and "avoiding intensive forest operations," and as a result has maintained an old-growth forest structure while still producing timber. As a result,

D. Fukui (✉)
The University of Tokyo Fuji Iyashinomori Woodland Study Center, Graduate School of Agricultural and Life Sciences, The University of Tokyo, Yamanakako, Japan
e-mail: fukuidai@uf.a.u-tokyo.ac.jp

S.N. Suzuki
The University of Tokyo Hokkaido Forest, Graduate School of Agricultural and Life Sciences, The University of Tokyo, Furano, Japan

Nakagawa Experimental Forest, Field Science Center for Northern Biosphere, Hokkaido University, Otoineppu, Japan
e-mail: snsuzuki@fsc.hokudai.ac.jp

T. Owari et al. (eds.), *Integrative Forest Management and Silviculture*,
Ecological Research Monographs, https://doi.org/10.1007/978-981-95-1017-7_15

it is a preferable habitat for forest organisms that originally inhabited Hokkaido and maintains high biodiversity (see Chaps. 5 and 6). In addition, the UTHF has established areas where no logging is allowed, such as forest reserves and restricted forests, which also contribute to the conservation of forest organisms. On the other hand, as in other areas in Japan, environmental change and the increase of nonnative species are becoming one of the threats to some species and ecosystems. This chapter introduces biodiversity and conservation and the management of nonnative species.

15.1 Reserved Forests and Restricted Forests

Relatively early in its history, the UTHF has maintained primary or near-primary forests by establishing forest reserves where no logging is allowed. The oldest conservation forest in existence is the Jinja-yama Forest Reserve, established in 1927 (Fig. 15.1). This may have been partly for religious reasons. Because "Jinja-yama" means "Shrine Mountain," and shrines have been erected on Jinja-yama since the beginning of the settlement. In Japan, it is customary to prohibit logging in the forests around shrines as sacred forests. Subsequently, characteristic forests in various locations were designated as conservation forests (refer to Chap. 5). However, by the 1980s, all of them were only a few ha or several tens of ha in area.

Fig. 15.1 Jinja-yama Forest Reserve. (Photo: Makoto Ishiguro, 2023.9.11. Used with permission from the UTHF, licensed under CC BY-NC-ND 4.0)

Fig. 15.2 Nishitappu-Oku Primeval Forest Reserve. (Photo: Ishiguro Makoto, 2023.10.16. Used with permission from the UTHF, licensed under CC BY-NC-ND 4.0)

In 1972, a large 1240 ha Maeyama Forest Reserve was established to preserve most of the southwestern slopes (elevation 550–1459 m) of Mt. Dairoku. In this area, light selection cutting was conducted once or twice by the 1937, but has never been logged since then, maintaining a near-primary condition. The Nishitappu-Oku Primeval Forest Reserve, established in 1932, is relatively large (46 ha) and is also a conservation forest with no logging history (Fig. 15.2). Although the area was severely affected by the typhoon in 1981, it is recognized for preserving the old-growth primeval forests of the region. The area of this reserve is designated as the Special Protection Area of Wildlife Protection Area by Hokkaido government (see 15.3). Unfortunately, there are no permanent plots for continuous tree measurements and research on its dynamics has not been conducted. Other reserves include the Iwanazawa Forest Reserve (93 ha), which protect riparian forests, the Tairazawa Forest Reserve (36 ha), dedicated to lowland wet forests, and the Sannosawa Jobu Forest Reserve, conserving natural Sakhalin spruce wetland forests, among others. As described in Chap. 5, areas with specially unique plant communities, such as those around the blowing caves and limestone vegetation, are designated as a Special Plant Community Protection Area.

Ecosystem Conservation Forests and Difficult-to-Operate Forests (refer to Chap. 8) may also contribute to biodiversity conservation. Ecosystem Conservation Forests are designated to conserve areas with high vulnerability, such as riparian forests, wetlands, alpine zones, and wind-exposed areas. Difficult-to-Operate Forests are areas that will not be logged due to steep slopes or other factors that

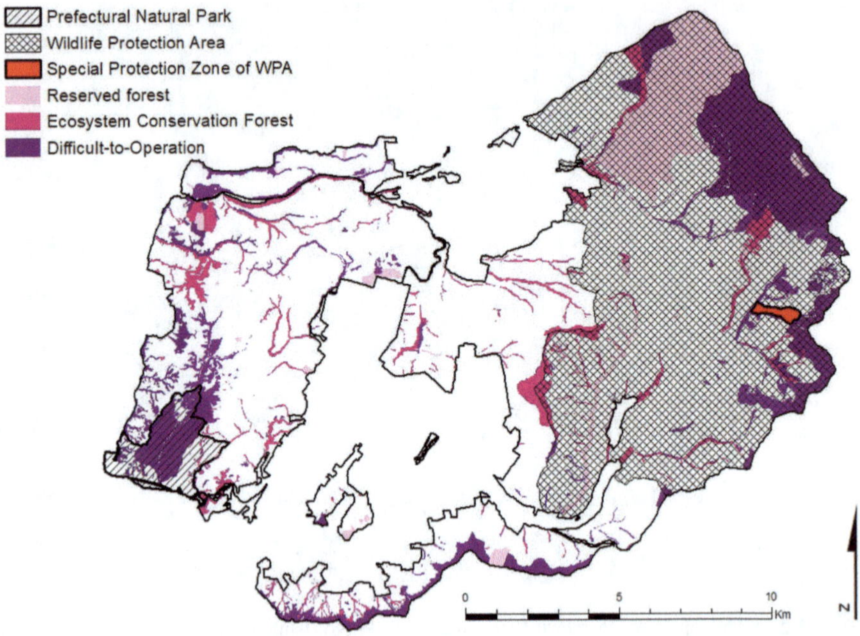

Fig. 15.3 The forest reserves and special plant community protection areas, Ecosystem Conservation Forest, Difficult-to-Operation Forest, Wildlife Protection Area, Prefectural Natural Park in UTHF. (Created by Shinya Inukai after The University of Tokyo Forests (2022). Used with permission from the UTHF, licensed under CC BY-NC-ND 4.0)

make access by heavy machinery difficult. Although these forests are not specifically intended for biodiversity conservation, logging has been restricted in them, providing refuges for forest organisms and helping to maintain populations of old, large trees and dead wood.

At the beginning of the 14th period, 1484 ha were in Forest Reserves and Special Plant Community Protection Area, and 4521 ha in Ecosystem Conservation Forests and Restricted Forests, leaving approximately 6000 ha (26% of UTHF) of forests unlogged (Fig. 15.3, The University of Tokyo Forests 2022).

Even in forests where logging is being conducted, the SSMS takes biodiversity conservation into consideration in its operations. For example, old trees that are no longer valuable as timber are left as nesting or foraging trees. In particular, trees that have cavities play a major role as nesting trees for owls and small mammals (Fig. 15.4a). Tall trees with branches on ridges are also important as perches for raptors and should be preserved whenever possible (Fig. 15.4b). Dead and fallen trees will also be conserved whenever possible as a resource and habitat for dead and dying tree-dependent organisms as well as nursery logs for tree regenerations (Okada et al. 2019).

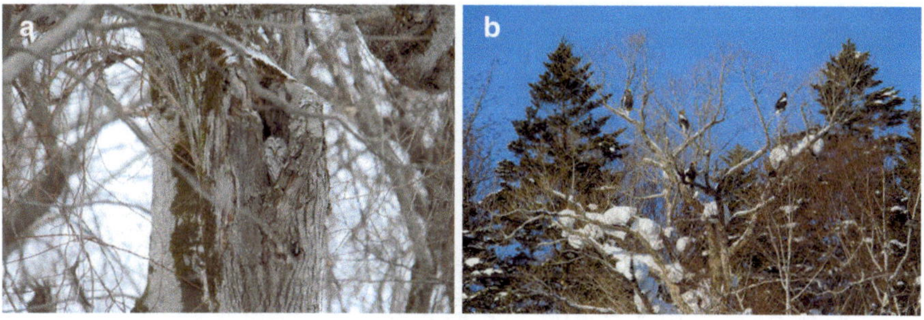

Fig. 15.4 (**a**) Ural owl (*Strix uralensis japonica*) in tree cavity. (**b**) Steller's sea eagles (*Haliaeetus pelagicus*) on branches of a large tree. (Photo by Nozomi Oikawa, used with permission from the UTHF, licensed under CC BY-NC-ND 4.0)

15.2 Wildlife Protection Area and Natural Park

The eastern half of the UTHF (11,059 ha) has been designated as a Wildlife Protection Area by the Hokkaido government under the Wildlife Protection, Control, and Hunting Management Act, where game hunting is prohibited. In addition, 45 ha of this area (Nishitappu-Oku Primeval Forest Reserve) is designated as a Special Protection Zone. Another 654 ha, the southwestern part of the UTHF, has been designated as a Prefectural Natural Park of Hokkaido Prefecture. Game hunting is also prohibited in this area.

15.3 Rare Species

Numerous rare plants and animals are existing in the UTHF. So far, 894 taxa of vascular plants have been identified in the UTHF (see Chap. 5), of which 40 species are on the Red List of the Ministry of the Environment (Horie et al. 2013). For birds and mammals, 12 and three species are on the Red List of the Ministry of the Environment, respectively (Tables 15.1 and 15.2).

15.4 Impacts Against Increasing Animals, Environmental Changes, and Nonnative Species

*15.4.1 Sika Deer (*Cervus nippon*)*

As introduced in Chap. 14, the sika deer population that have begun to increase since the 1990s has had a significant impact on the ecosystem. The impact is not so severe at present, but it is pronounced in some species, regions, and time periods. Although the nursery and some sample forests areas in the arboretum have been

Table 15.1 List of endangered birds in the UTHF designated by the Ministry of Environment. This list includes data deficient (DD) species

Rank	Order	Family	Genus	Species (subspecies)	Common name	Remarks
DD	Galliformes	Phasianidae	*Tetrastes*	*bonasia vicinitas*	Hazel grouse	
DD	Anseriformes	Anatidae	*Aix*	*galericulata*	Mandarin duck	
NT	Gruiformes	Rallidae	*Porzana*	*fusca erythrothorax*	Ruddy-breasted crake	
NT	Caprimulgiformes	Caprimulgidae	*Caprimulgus*	*indicus jotaka*	Jungle nightjar	
NT	Charadriiformes	Scolopacidae	*Gallinago*	*hardwickii*	Latham's snipe	
NT	Accipitriformes	Pandionidae	*Pandion*	*haliaetus haliaetus*	Western osprey	
VU	Accipitriformes	Accipitridae	*Haliaeetus*	*albicilla albicilla*	White-tailed eagle	Natural monument
VU	Accipitriformes	Accipitridae	*Haliaeetus*	*pelagicus*	Steller's sea eagle	Natural monument
NT	Accipitriformes	Accipitridae	*Accipiter*	*nisus nisosimilis*	Eurasian sparrowhawk	
NT	Accipitriformes	Accipitridae	*Accipiter*	*gentilis fujiyamae*	Northern goshawk	
EN	Accipitriformes	Accipitridae	*Nisaetus*	*nipalensis orientalis*	Mountain hawk-eagle	
VU	Piciformes	Picidae	*Dryocopus*	*martius martius*	Black woodpecker	Natural monument
VU	Falconiformes	Falconidae	*Falco*	*peregrinus japonensis*	Peregrine falcon	

Table 15.2 List of endangered mammals in the UTHF designated by the Ministry of Environment

Rank	Order	Family	Genus	Species	Common name
NT	Lagomorpha	Ochotonidae	*Ochotona*	*hyperborea*	Northern pika
NT	Carnivora	Mustelidae	*Mustela*	*erminea*	Ermine, stoat
NT	Carnivora	Mustelidae	*Martes*	*zibellina*	Sable

Fig. 15.5 Manchurian Elm with its bark eaten by sika deer. (Photo by the UTHF, licensed under CC BY-NC-ND 4.0)

prevented from being damaged by installing deer-proof fences, in the vicinity of the arboretum, most of the shrubs have disappeared due to very high densities of sika deer during the winter and snowmelt seasons. As described in Chaps. 5 and 14, some tree species, such as *Ulmus laciniata* and *Hydrangea paniculata*, have declined catastrophically in the Iwanazawa riparian forest (Fig. 15.5). In the UTHF, except for the Wildlife Protection Area, hunting by general hunters is conducted from October to March (varies depending on the year). The fundamental damage prevention measure is to keep the deer density low by hunting, etc. However, considering the population fluctuations in the surrounding areas, it seems necessary to apply more pressure to capture them.

15.4.2 Climate Change

As described in Chap. 4, the average temperature in the UTHF has risen by about 1.2 degrees since 1990 (Tanaka et al. 2013). This increase in temperature could have serious effects, especially in high mountain areas. In the UTHF, the upper part of Mt. Dairoku, between 1250 m and 1459 m a.s.l., is an alpine ecosystem, and some organisms are only existing in this area. For example, according to Oikawa and Matsui (2009), it is known that the Japanese pika (*Ochotona hyperborea*) only lives above 1340 m a.s.l. in the upper part of Mt. Dairoku. The Japanese pika's habitat is rocky masses with small crevices, and these crevices provide the preferable temperature environment (Onoyama, 1991). Rising temperatures could change forest structure and alter current habitat conditions, potentially impacting Japanese pika both directly and indirectly. Thus, if climate change were to further reduce alpine ranges, these alpine organisms may no longer be able to maintain viable population sizes.

15.4.3 Nonnative Species

Raccoons (*Procyon lotor*) are native to the North American continent, but were confirmed to have become established in the wild in Japan in 1962 and in Hokkaido in 1979. In Furano City, where the UTHF is located, it has been confirmed that this species has become established in agricultural areas since 2000 but had not been living within the forest for a long time. However, as part of camera trap monitoring of medium to large mammals that began in 2011 (see Chap. 6), this species was photographed for the first time in the arboretum in 2014. Since then, they have been observed almost every year in other areas of the UTHF, and it is speculated that they are establishing and increasing in number within the forest (Fig. 15.6). So far, no negative impacts on the forest ecosystem have been confirmed, but it is necessary to continue to monitor them.

Black locust (*Robinia pseudoacacia*: Fabaceae) are native to the North American continent and introduced to Japan as park trees in 1873. It had been widely used as street tree and a plant for erosion control for roadsides and river sides as it is fast growing and generally stress tolerant. This species is now naturalized across Japan and invaded into open sites such as riverbeds and vacant lands, and used as an important source plant of nectar. In the UTHF, black locust is only found in river side and forest edge, and rarely found in inner forest so far (Fig. 15.7). However, it can potentially invade into the site where forest stands have been recently clear-cut because of its fast growth rate.

A number of alien herbaceous plants have been invaded into open sites such as the roadsides and newly planted sites in the UTHF (Fig. 15.8). *Rudbeckia laciniata* (Asteraceae), which is native to the North American continent, is one of the problematic invasive species in Hokkaido, and is listed as a species of the Designated Invasive Alien Species by the Ministry of Environments. *Solidago gigantea* var.

Fig. 15.6 A raccoon taken by a camera trap. (Photo by the UTHF, licensed under CC BY-NC-ND 4.0)

Fig. 15.7 A *Robinia pseudoacacia* community along a river. (Photo by Nozomi Oikawa, used with permission from the UTHF, licensed under CC BY-NC-ND 4.0)

Fig. 15.8 (**a**) *Rudbeckia laciniata* community, (**b**) *Solidago gigantea* var. *leiophylla* community, (**c**) *Cirsium vulgare*, and (**d**) *Conyza canadensis* community. You can also see the individuals of *C. vulgare* and *Conyza* sp. in (**d**). (Photo by Nozomi Oikawa (**a, b**), the UTHF (**c**), and Satoshi N. Suzuki (**d**), used with permission from the UTHF, licensed under CC BY-NC-ND 4.0)

leiophylla, *S. altissima*, *Cirsium vulgare*, *Conyza canadensis*, all of them are also Asteraceae and native to the North American continent, are also wide spreading invasive species in Hokkaido, and often cover densely in open sites. Because these species become generally taller than 1 m and often reaches 2 m in late summer to autumn, these species become a competitor for the planted tree seedlings and prevent trees from natural regeneration in plantations. Fortunately, invasive species are rarely found in inner of forests, even after selection cutting. Therefore, the selection cutting system can be robust to such invasive plant species, at least in Hokkaido.

Exotic species planted in plantations can be sources of alien species spread and genetic contamination. Eastern white pine (*Pinus strobus*), native to North America, often regenerates naturally after scarification. Takahashi et al. (2023) found that *P. strobus* regenerated in scarified areas after clear-cutting but did not in nonscarified areas. Exotic white birch species, such as *Betula papyrifera*, have been planted in trials and may potentially regenerate in open sites. Additionally, these exotic white birch species may hybridize with the native white birch (*B. platyphilla*). Norway spruce (*Picea babies*) can also hybridize with native spruces (*Picea glehnii*). Such hybridization between exotic and native species can lead to genetic disturbance if the hybrids can backcross with native species. Although genetic contamination of native species has not been examined in the UTHF, it may become an issue in the future. Since the UTHF does not plan to plant any more nonnative species, the risk of genetic disturbance is expected to decrease gradually.

References

Horie K, Miyamoto Y, Kimura N, Oikawa N (2013) List of vascular plant of the University of Tokyo Hokkaido Forest. Misc Info Univ Tokyo Forests 54:59–106

Oikawa N, Matsui M (2009) Tokyo Daigaku Hokkaido Ensyurin ni okeru Ezo-nakiusagi no Seisoku Joukyou [habitat status of the northern Pika in the University of Tokyo Hokkaido Forest]. Northern Forestry Japan 61:11–14

Okada M, Hirao T, Kaji M, Goto S (2019) Role of fallen logs in maintaining the species diversity of understory vascular plants in a mixed coniferous and broad-leaved forest in Hokkaido, northern Japan. For Ecol Manag 448:249–255

Onoyama K (1991) Nisshu Katsudou [Circudian activity] in: division of natural environment, department of health and environment. In: Hokkaido (ed) Field study report such as wild animal distribution. Surveys reports such as Ochotona hyperborea ecology. Division of Natural Environment, Department of Health and Environment, Hokkaido, Sapporo, pp 56–65

Takahashi K, Suzuki SN, Goto S (2023) Conversion of a *Pinus strobus* plantation to a conifer-broadleaved mixed forest by small-scale clear-cutting and scarification treatments: regeneration dynamics of *P. strobus* and native tree species during 15 years. J Japanese Forest Soc 105:65–75. (in Japanese with English summary)

Tanaka N, Kamata N, Shibano H, Owari T, Ohkawa A, Igarashi Y, Arakida K (2013) Estimation of long-term data sets of annual mean temperature in Ecohydrology research institute, the University of Tokyo Hokkaido Forest, and the University of Tokyo Chichibu Forest. Bullet Tokyo Univ Forests 128:1–19

The University of Tokyo Forests (2022) Education and research plan (2021–2030) of the University of Tokyo Forests: part 3 regional Forest plans: the University of Tokyo Hokkaido Forest (the 14th period). Misc Info Univ Tokyo Forests 64:103–190

Chapter 16
Ecosystem Services in the UTHF

Satoshi N. Suzuki, Nobuaki Tanaka, and Toshiaki Owari

Abstract The University of Tokyo Hokkaido Forest (UTHF) contributes to a variety of ecosystem services, including carbon storage and sequestration, water supply, and cultural services. The estimated carbon absorption by living trees in the UTHF is approximately 7908 Mg-CO_2 y^{-1}, which accounts for 47% of the total emissions from Furano City. Remote sensing methods have also been employed to evaluate the carbon stocks. While several studies have examined the carbon stocks of components other than living trees, more comprehensive research is needed. Streamwater from the UTHF has been intensively utilized as both drinking and nondrinking water resources. A number of hydraulic infrastructures are installed along the UTHF streams, eventually providing 0.13 and 200 tons of water for drinking and nondrinking water, respectively, to surrounding communities and croplands. In addition to these services, the trees in the UTHF have also supported the traditional activities of the indigenous Ainu people. The forest provides recreational opportunities and serves a source of various wild vegetables.

According to the Millennium Ecosystem Assessment (2005) defined by the United Nations, ecosystem services are classified into four categories: (1) provisioning services, (2) regulating services, (3) cultural services, and (4) supporting services. While previous chapters have primarily focused on timber production and

Authors Satoshi N. Suzuki and Nobuaki Tanaka have equally contributed to this chapter.

S. N. Suzuki (✉)
The University of Tokyo Hokkaido Forest, Graduate School of Agricultural and Life Sciences, The University of Tokyo, Furano, Japan

Nakagawa Experimental Forest, Field Science Center for Northern Biosphere, Hokkaido University, Otoineppu, Japan
e-mail: snsuzuki@fsc.hokudai.ac.jp

N. Tanaka · T. Owari
The University of Tokyo Hokkaido Forest, Graduate School of Agricultural and Life Sciences, The University of Tokyo, Furano, Japan
e-mail: tnk-nobu@g.ecc.u-tokyo.ac.jp; owari@g.ecc.u-tokyo.ac.jp

T. Owari et al. (eds.), *Integrative Forest Management and Silviculture*,
Ecological Research Monographs, https://doi.org/10.1007/978-981-95-1017-7_16

biodiversity conservation, this chapter explores other ecosystem services provided by the UTHF, such as carbon storage and sequestration, water supply, and cultural services.

16.1 Carbon Storage and Sequestration

As concerns about rising atmospheric CO_2 levels and climate change grow, there is increasing interest in the carbon storage and absorption capabilities of forests. In the UTHF, there is anticipation for multifaceted functionality that encompasses not only timber production and biodiversity conservation but also carbon storage and sequestration.

16.1.1 Total Carbon Stock of Trees in the UTHF

The current total growing stock of trees in the UTHF is approximately 5.4 million m^3 (5,400,738 m^3 in 2021), based on measurements from over four thousand inventory plots (Management Coordinator, UTF 2023). This corresponds to a carbon stock of 2,126,678 Mg-C, according to the Japanese government's calculation method (National Institute for Environmental Studies, Japan 2021). The annual carbon absorption by the forest is estimated to be roughly 28,222 Mg-C y^{-1}, assuming an average growth rate for the region (2.76% $yr.^{-1}$ for plantation and 1.16% y^{-1} for natural forests). After accounting for tree logging (6975 Mg-C y^{-1}), the net annual carbon absorption is approximately 21,247 Mg-C y^{-1}, equivalent to 77,908 Mg-CO_2 y^{-1}. This amount accounts for 47% of the total emissions from the University of Tokyo in 2021 (The University of Tokyo 2023) and 47% of the total emissions from Furano city in 2021 (Ministry of the Environment of Japan 2024).

The above estimate of carbon sequestration is based on a simplified method. The carbon content of trees is calculated using average values for wood density, biomass expansion factors (BEFs), and root ratios for coniferous or broadleaf trees, rather than species-specific values. BEF and root ratios are assumed to be constant across tree sizes, although these values can vary depending on tree sizes. The timber volume per trunk is estimated via volume tables based solely on DBHs measured at 2 cm intervals, without considering tree height. The total forest accumulation at a given time was derived from the most recent inventory results of each forest compartment and the average growth rates for each forest stand type. Since these growth rates fluctuate due to various factors, they can also introduce uncertainty into the estimate. Moreover, it is essential to verify how well the inventory results from sample plots represent the overall forest.

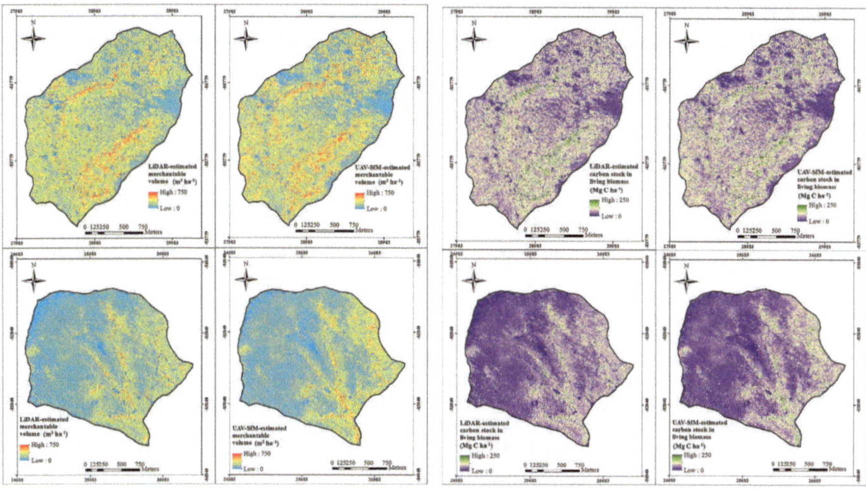

Fig. 16.1 Spatial distribution of the LiDAR-estimated (the left two panels in both the left and right groups of four panels) and UAV–SfM-estimated (the right two panels in both the left and right groups of four panels) merchantable volume (left four panels) and carbon stock (right four panels) in the compartments 43 (upper panels) and 48 (lower panels) in the UTHF (50 m pixel resolution). (Reproduced from Jayathunga et al. 2018, with the publisher's permission)

16.1.2 Remote-Sensing Approach

Methods for estimating forest carbon stocks using remote-sensing techniques have been advanced significantly over the past decade. The UTHF has also utilized these methods to estimate carbon stocks. Jayathunga et al. (2018) compared the effectiveness of airborne LiDAR and Structure-from-Motion (SfM) analysis of aerial photos taken by UAVs in estimating the merchantable volume and carbon stock of forest stands (Fig. 16.1). The authors found that SfM-UAV methods provide highly accurate estimates of stand carbon stocks, comparable to those obtained using the airborne LiDAR method. This finding suggests that SfM-UAV approach can be a cost-effective means for estimating carbon stocks at the compartment level (less than 200 ha).

In the UTHF, stand parameters, including the carbon stock, were estimated using UAV photogrammetry in a 115-year-old Japanese larch (*Larix kaempferi*) plantation (Karthigesu et al. 2023) and in a mixed conifer-broadleaf forest (Karthigesu et al. 2024). The UAV's flying altitude and filtering methods greatly impacted the accuracy of stand parameter estimation in old-aged plantations, and UAV flights at 120 m yielded higher accuracy for carbon stock compared to flights at 80 m when using Gaussian and mean filtering. Additionally, the combined use of structural, textural, and spectral metrics derived from UAV RGB imagery with random forest and multiple linear regression models effectively improved the estimation of stand parameters in a mixed conifer-broadleaf forest.

Table 16.1 Carbon storage (t $ha^{-1}C$) in trees and Sasa biomass in stands with different silviculture treatments 25 years after the stand replacement disturbance caused by the typhoon in 1981. "Level" means the depth of soil scarification for the scarification treatment, and planted species for the plantation treatment: *Lgk*, *Larix gmelinii* var. *japonica* ×*kaempferi*; *Pg*, *Picea glehnii*; *Pj*, *Picea jezoensis*; *As*, *Abies sachalinensis* (after Owari et al. 2011)

		Tree		Sasa		Total
Treatment	Level	Mean	SE	Mean	SE	Mean
Scarification	0 cm	51.0	6.3	13.7	17	64.7
	10 cm	29.0	6.8	9.8	1.4	38.7
	20 cm	24.8	n.a.	15.5	1.8	40.3
Plantation	*Lgk*	72.6	3.4	7.8	1.6	80.3
	Pg	58.6	6.2	9.1	0.9	67.7
	Pj	37.2	3.8	13.3	2.0	50.4
	As	19.0	7.8	21.5	2.3	40.5
Control[a]		21.1 (42.5)	2.9 (5.5)	14.4	1.9	35.5 (42.5)

[a]Figures in parentheses are carbon storage in advanced regeneration tree biomass

16.1.3 Carbon Stock of Components Other Than Living Trees

In boreal forests, components other than living trees, such as understory vegetation, fine litter, coarse woody debris, and soil organic matter, play a significant role in carbon stocks, as living trees account for only 10–35% of the total carbon stock (DeLuca and Boisvenue 2012). In the UTHF, however, the carbon stock of these components has not been systematically examined. One notable study by Owari et al. (2011) estimated the carbon stock of not only trees but also Sasa dwarf bamboos in forest stands under a variety of treatments after a typhoon disturbance in the UTHF. They found that the carbon stock of Sasa dwarf bamboos accounted for 13–40% of the total vegetation carbon stock in stands 25 years after the typhoon disturbance. Additionally, they evaluated carbon emissions from posttyphoon silviculture operations, such as soil scarification and planting, reporting that these emissions from silviculture operations were much lower than the carbon sequestered in afforested stands (Table 16.1).

A comprehensive study on the carbon balance following large-scale wind disturbance was conducted by Hotta et al. (2023). They simulated 115 years (2015–2130) of forest dynamics under various scenarios, involving different climate changes, windthrow regimes, and management methods using the LANDIS-II forest landscape model. CO_2 emissions related to management and salvaged wood were also estimated through a life cycle assessment (LCA) approach (Fig. 16.2a). The study reported that the 115-year cumulative net CO_2 absorption of the forest sector, including the carbon balance within ecosystems and CO_2 emissions estimated by LCA, significantly decreased due to salvage logging (up to 81%) and scarification (up to 114%) (Fig. 16.2b).

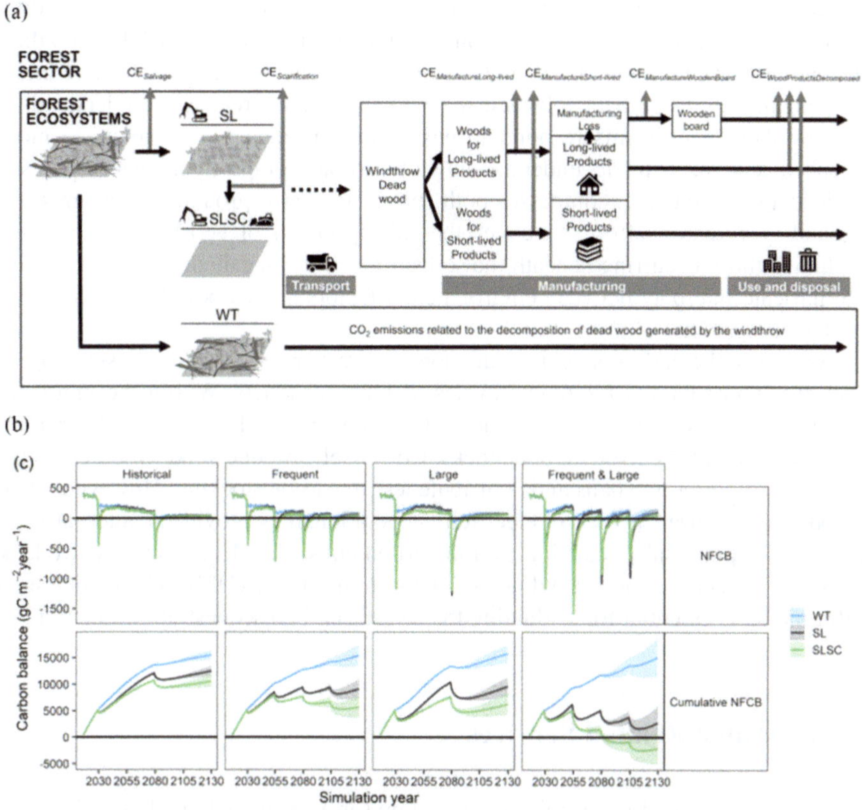

Fig. 16.2 (**a**) Schematic diagram of the life cycle assessment in Hotta et al. (2023). Gray arrows indicate CO_2 emissions by each factor. Dark gray boxes indicate stages in the life cycle. The part indicated by a dotted arrow was cut off in the life cycle assessment. *WT* windthrow only, *SL* salvage logging, *SLSC* salvage logging, and scarification. $CE_{Salvage}$: CO_2 emissions related to salvage logging (e.g., use of heavy machinery); $CE_{Scarification}$: CO_2 emissions related to scarification (e.g., use of heavy machinery); $CE_{ManufactureLong\text{-}lived}$: CO_2 emissions related to manufacturing of long-lived products (timber products used for several decades such as building materials and furniture); $CE_{ManufactureShort\text{-}lived}$: CO_2 emissions related to manufacturing of short-lived products (products to be disposed of in a few years such as wood chips and paper); $CE_{ManufactureWoodenBoard}$: CO_2 emissions related to the manufacturing of wooden boards (such as particle boards); $CE_{WoodProductsDecomposed}$: CO_2 emissions related to the disposal of products that were made from dead wood generated by windthrow. (**b**) Chronological changes in the net forest sector carbon balance (NFCB) and the cumulative NFCB under four scenarios with different windthrow regimes (historical regimes, more frequent disturbances, larger disturbances, and more frequent and larger disturbances than historical regimes) with the three different treatments after windthrow (*WT*, *SL*, and SLSC). In terms of the vertical axis, negative values indicate net CO_2 sources, while positive values indicate net CO_2 sinks. The solid lines and shaded areas indicate the mean values and standard deviations within the climate scenarios, respectively. (Reproduced from Hotta et al. 2023, with the publisher's permission)

Observational data on carbon stocks in belowground components, including coarse woody debris, fine litter, and soil organic matter, as well as insights into their decomposition processes, are notably scarce within the UTHF. This lack of information presents a significant obstacle to developing a comprehensive understanding of how these ecosystems respond to environmental changes. Achieving a more nuanced and mechanistic understanding of carbon storage dynamics requires a holistic approach that encompasses both above- and belowground components. By expanding research efforts to cover the full spectrum of ecosystem processes, including those occurring beneath the soil surface, we can gain deeper insights into the intricate interplay between environmental factors and carbon dynamics within UTHF ecosystems.

Assessing the carbon stocking function of forests requires an understanding of both the industrial and biological cycles of forest products. Within the industrial cycle, woody products such as structural components of buildings and furniture serve as crucial long-lived carbon stocks. Conversely, products like wood chips for paper have shorter lifespans and contribute less to carbon stocking. However, when wood chips are used as fuel for electricity generation, they provide an alternative to fossil fuels, potentially reducing overall carbon emissions. Unfortunately, we lack quantitative data on the breakdown of wood usage from UTHF. Collecting such information is essential for evaluating the forest's carbon sequestration function.

16.2 Water-Related Services

There are a variety of water-related ecosystem services, including (1) provisioning water for both drinking and nondrinking purposes, (2) regulating water quality, soil erosion, and flooding, (3) providing cultural services such as recreation, and (4) supporting hydropower generation and sand gravel extraction. In the case of the UTHF, the most important water-related service is provisioning, as UTHF streamwater is intensively utilized for both drinking and nondrinking water.

The Furano municipal government currently operates a central purification plant along the Upper Sorachi River to supply drinking water primarily to the downtown (Furano City 2021). To provide drinking water to suburban areas, the government also operates six simplified water purification stations (SSs) (Furano City 2021). Additionally, in remote areas not managed by the government, 22 community-based associations (CAs) are responsible for drinking water management and supply to their respective communities. Since the UTHF is located upstream of these areas, two SSs and eight intake infrastructures for CA are installed along small streams within the UTHF (Fig. 16.3), providing an annual total of 0.13 million tons of drinking water to surrounding areas.

Since the Furano Basin (Fig. 3.1b) was originally a peatland, large-scale drainage and irrigation projects have been undertaken since the period of settlement,

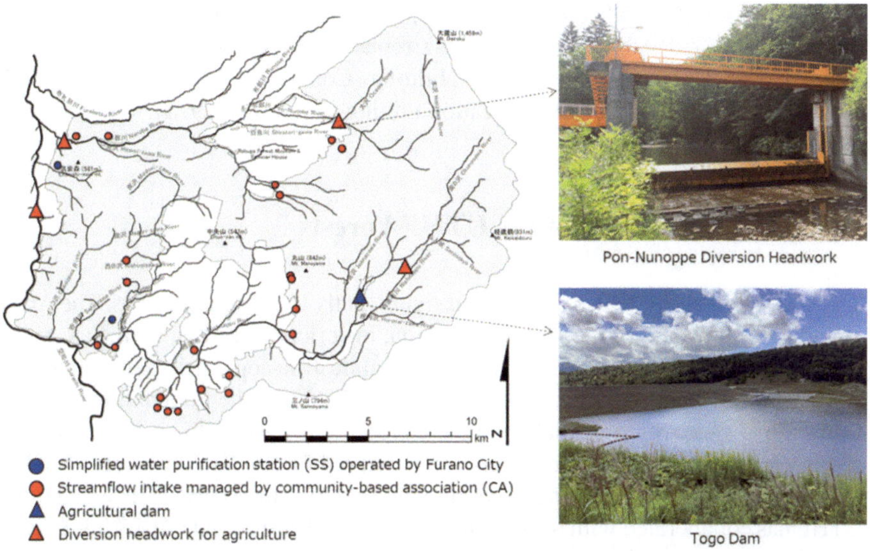

Fig. 16.3 Map of hydraulic infrastructures in the UTHF (shade area). Basemap is provided by UTHF with permission and locations of SS were taken from Furano City (2021). (Photos by Nobuaki Tanaka)

enabling agricultural activities to flourish. The earliest drainage project began in 1917, initiated by Mr. Tokuhei Kabutoya, a previous village council member. He established a cooperative for improving the soil, and over a period of two years, the cooperative completed the construction of a 4.5-km artificial drainage canal running north–south through the Furano Basin, facilitating the cultivation of approximately 3000 hectares of peatland. However, due to the high sulfate concentration of the Furano River (Nishi et al. 1974, see also Chap. 4), which originates from the active volcano Mount Tokachi, it cannot be utilized for irrigation. Therefore, in 1921, Kabutoya reorganized the cooperative and constructed a 22.3-km irrigation canal along the eastern edge of the Furano Basin to obtain irrigation water from the Upper Sorachi River. Since the 1970s, national irrigation and drainage projects have been extensively implemented. As a result, one agricultural dam and three diversion headworks in (or adjacent to) UTHF streams are currently in operation, providing an annual total of 200 million tons of water resources to approximately 8000 hectares of paddy fields and croplands (Furano Land Improvement District 2024). In general, timber harvesting and forest management practices can alter streamflow in terms of both quantity and quality. Consequently, the UTHF is currently investigating whether there is any trade-off relationship between timber harvesting by the stand-based silvicultural management system and the water-related ecosystem services of the UTHF.

Both the Nishitappu and Nunobe Rivers were utilized to generate hydropower from the 1930s to the 1940s, supporting the early electrification of the UTHF facilities, including headquarters and staff quarters. The surplus electricity was distributed to nearby villagers and farmers. This early electrification ceased during the 1950s to the 1960s, eventually being replaced by regional-scale commercial power (Higashiyama Local History Editing Committee 1989; Rokugo 50th Anniversary Commemorative Publication Editing Committee 1971).

16.3 Other Services of the UTHF Forests

The UTHF forests have also contributed significantly to cultural services. The Ainu, an indigenous people of the Hokkaido region, historically used large trees to make dugout canoes. However, this technique has almost disappeared even among the Ainu people. To preserve this traditional craft and culture, the Foundation for Ainu Culture has endeavored to make dugout canoes themselves. However, trees large enough for this purpose are found only in preserved forests and are rarely found in managed forests where logging is permitted, except for the UTHF. Since 1998, the UTHF has collaborated with the Foundation for Ainu Culture and other Ainu-related organizations, providing approximately 20 large trees such as Katsura trees (*Cercidiphyllum japonicum*), castor aralia (*Kalopanax septemlobus*). The dugout canoes made from the UTHF tree have been used and exhibited at the Upopoy National Ainu Museum and Park (Fig. 16.4a).

Fig. 16.4 The dugout canoe in the Upopoy National Ainu Museum and Park (**a**, photo by the UTHF, June 30, 2022, licensed under CC BY-NC-ND 4.0), and the wine barrel previously used in a local winery (**b**, photo by Nobuaki Tanaka, August 8, 2024)

Woods from the UTHF play a significant role in supporting local cultural and educational initiatives. In 2021, the city Furano launched a "wood education project," which involves gifting wooden building blocks, called "Furano-no-Tsumiki," made from UTHF timber, to newborn children in the city. This project not only provides opportunities for children to become familiar with the local natural resources of their region from young age but also fosters a connection with the local environment and the importance of sustainable forestry practices. Additionally, for the 50th anniversary of a prominent local winery, the UTHF provided Japanese oak logs specifically for the production of wine barrels (Fig. 16.4b). These collaboration between the UTHF and local communities underscores the forest's role in supporting local traditions and economy, while also promoting sustainable use of forest resources.

UTHF also allows the public to enjoy its majestic nature. In collaboration with Furano city, the UTHF hosts several public events and offers guided tours through its forest. Participants are often captivated by the sight of an enormous oak tree or the stunning view from the summit of Mt. Dairoku (Figs. 16.5 and 16.6). Visitors can also experience the beautiful scenery, refreshing cool air, and the soothing sounds of water in the Nishitappu Spring (refer to Chap. 4).

Fig. 16.5 Symbolic large Japanese oak tree in Jinja-yama. (Photo by the UTHF, May 9, 2017, licensed under CC BY-NC-ND 4.0)

Fig. 16.6 View from the summit of Mt. Dairoku. (Photo by the UTHF, Jul. 30, 2008, licensed under CC BY-NC-ND 4.0)

Although not accessible to the public, the UTHF has the potential to provide a variety of wild vegetables. In spring, alpine leek (*Allium victorialis*), an edible and popular wild vegetable in Hokkaido, sprouts up in various areas. At high-altitudes, it is also possible to harvest edible bamboo shoots from *Sasa kurilensis*. The spring shoots of certain trees, such as Japanese angelica tree (*Aralia elata*), castor aralia, and Koshiabura tree (*Chengiopanax sciadophylloides*), are also edible and enjoyed as wild vegetables. Additionally, a variety of mushrooms emerge on forest floors, dead wood materials, and living stems, especially in autumn.

Fishing is also restricted in UTHF, allowing it to serve as a refuge for fish such as white-spotted char (*Salvelinus leucomaenis leucomaenis*), a popular native target of fishing. This protection can contribute to the sustainability of fishing for this species in the region.

16.4 Remarks

The UTHF serves as a valuable natural resource that provides a wide range of ecosystem services. It is essential not only for the local community but also for maintain, supporting their well-being as well. Ongoing research and collaboration with local communities are essential for advancing our understanding of forest ecosystems and enhancing the sustainable use of forest resources. This comprehensive range of benefits highlights the importance of preserving and managing UTHF carefully to ensure it continues to provide valuable services for future generations.

References

Deluca TH, Boisvenue C (2012) Boreal forest soil carbon: distribution, function and modelling. Forestry Int J Forest Res 85:161–184. https://doi.org/10.1093/forestry/cps003

Furano City (2021) Furano City water supply vision

Furano Land Improvement District (2024) Daichi, Newsletter of Furano Land Improvement District, vol 48, p 12

Higashiyama Local History Editing Committee (Ed.) (1989) *Higashiyama Kyoudoshi* [Higashiyama local history], Higashiyama 80th anniversary commemorative project executive Committee, Japan, 594pp

Hotta W, Haga C, Morimoto J, Suzuki SN, Matsui T, Owari T, Shibata H, Nakamura F (2023) Leaving disturbance legacies conserves boreal conifers and maximizes net CO2 absorption under climate change and more frequent and larger windthrow regimes. Landsc Ecol 38:1785–1805

Jayathunga S, Owari T, Tsuyuki S (2018) The use of fixed–wing UAV photogrammetry with LiDAR DTM to estimate merchantable volume and carbon stock in living biomass over a mixed conifer–broadleaf forest. Int J Appl Earth Obs Geoinformation 73:767–777

Karthigesu J, Owari T, Tsuyuki S, Hiroshima T (2023) UAV photogrammetry for estimating stand parameters of an old Japanese larch plantation using different filtering methods at two flight altitudes. Sensors 23(24):9907

Karthigesu J, Owari T, Tsuyuki S, Hiroshima T (2024) Improving the estimation of structural parameters of a mixed conifer-broadleaf forest using structural, textural, and spectral metrics derived from UAV RGB imagery. Remote Sens 16:1783. https://doi.org/10.3390/rs16101783

National Institute for Environmental Studies, Japan (2021) National greenhouse gas inventory report of Japan 2020. National Institute for Environmental Studies, Japan

Management Coordinator, Forest Management Committee, The University of Tokyo Forests, Graduate School of Agricultural and Life Sciences, The University of Tokyo (Management Coordinator, UTF) (2023) Calculation of CO2 sequestration by trees in the University of Tokyo Forests for "the University of Tokyo environmental report 2022". Miscell Info University Tokyo Forests 67:59–69

Millennium Ecosystem Assessment (2005) A report of the millennium ecosystem assessment. In: Ecosystems and human well-being. Island Press, Washington DC

Ministry of the Environment of Japan (2024) Jichitai haisyuturyo karute [charts for emissions of local governments], ministry of the environment, https://www.env.go.jp/policy/local_keikaku/tools/karte.html. Last Accessed 20th May 2024

Nishi T, Sugimoto K, Takahashi M, Sakai T, Yonekawa M, Miyagacho S, Kato G (1974) On the water examinations of the rivers at the piedmont district of Mt. Tokachi in and around Kamifurano Cho. Res Bull Obihiro Univ 8:319–343

Owari T, Kamata N, Tange T, Kaji M, Shimomura A (2011) Effects of silviculture treatments in a hurricane-damaged forest on carbon storage and emissions in Central Hokkaido, Japan. J For Res 22:13–20

Rokugo 50th Anniversary Commemorative Publication Editing Committee (Ed.) (1971) *Rokugo no 50 Nen* [Rokugo 50-year history], Rokugo 50th anniversary cooperation, Japan, 291pp

The University of Tokyo (2023) Environmental report 2023. The University of Tokyo. (In Japanese with English digest)

Part V
The Future

Chapter 17
Future Challenges and Opportunities for Integrative Forest Management and Silviculture

Toshiaki Owari, Satoshi N. Suzuki, Nobuaki Tanaka, and Dai Fukui

Abstract Throughout this book, the authors have examined the integrative forest management and silvicultural practices developed and implemented at the University of Tokyo Hokkaido Forest (UTHF) over the past 60 years. At the core of these efforts is the Stand-based Silvicultural Management System (SSMS), a pioneering approach that harmonizes timber production with biodiversity conservation and ecosystem health. By showcasing the SSMS as a model for integrative forest management and silviculture, we aim to inspire its adoption in forest ecosystems worldwide. This chapter concludes the discussion by summarizing the challenges and opportunities for future research at the UTHF, with a focus on advancing sustainable practices and enhancing the forest's ecological and social contributions.

Keywords Climate change impact · Deer overpopulation · Forest health and resilience · Natural regeneration · Technological advancement

T. Owari (✉) · N. Tanaka
The University of Tokyo Hokkaido Forest, Graduate School of Agricultural and Life Sciences, The University of Tokyo, Furano, Japan
e-mail: owari@g.ecc.u-tokyo.ac.jp; tnk-nobu@g.ecc.u-tokyo.ac.jp

S. N. Suzuki
The University of Tokyo Hokkaido Forest, Graduate School of Agricultural and Life Sciences, The University of Tokyo, Furano, Japan

Nakagawa Experimental Forest, Field Science Center for Northern Biosphere, Hokkaido University, Otoineppu, Japan
e-mail: snsuzuki@fsc.hokudai.ac.jp

D. Fukui
The University of Tokyo Fuji Iyashinomori Woodland Study Center, Graduate School of Agricultural and Life Sciences, The University of Tokyo, Yamanakako, Japan
e-mail: fukuidai@uf.a.u-tokyo.ac.jp

T. Owari et al. (eds.), *Integrative Forest Management and Silviculture*, Ecological Research Monographs, https://doi.org/10.1007/978-981-95-1017-7_17

17.1 Introduction

This book has examined the integrative forest management and silvicultural practices developed and implemented at the University of Tokyo Hokkaido Forest (UTHF) over the past 60 years. Central to these efforts is the Stand-based Silvicultural Management System (SSMS), a pioneering approach designed to balance timber production with biodiversity conservation and ecosystem health.

Part I provided an overview of the background and contents of this book, along with the historical evolution of forest management at the UTHF since its establishment in 1899. Key transitions included the early colonial forestry efforts, the adoption of the SSMS, and the integration of digital technologies. Significant milestones involved recovery from natural disasters such as fires and typhoons, responding to wartime resource demands, and the eventual shift toward sustainable forestry practices.

Part II explored the environmental complexities of the UTHF's hemi-boreal forest ecosystem. This section examined the forest's topography, geology, soil types, climate, and hydrology, offering a detailed understanding of how these factors shape management decisions. It also emphasized the rich biodiversity within the UTHF, focusing on long-term studies of vegetation and wildlife that provide valuable insights into forest dynamics and species interactions.

Part III focused on the principles and practical applications of the SSMS. It outlined six foundational principles, such as minimal intervention, adaptive management, and fostering diverse, multilayered forests. Key topics included:

- **Silvicultural Planning:** Processes like forest type classification, plot inventories, tree marking, and harvesting.
- **Harvesting Operations:** Timber extraction techniques, marketing strategies (e.g., auctioning high-value hardwoods), and forest road construction.
- **Assisted Natural Regeneration (ANR):** Transitioning from monoculture plantations to resilient, mixed-species forests.
- **Seedling Production:** Ensuring the production of high-quality seedlings to support natural regeneration and maintain genetic diversity.

Part IV shifted focus to optimizing multiple ecosystem functions at a landscape scale. This section addressed the effects of selection cutting on ecosystem processes, such as deadwood dynamics and the regeneration of *Picea jezoensis*. It also highlighted challenges posed by natural disturbances and biological threats, including pests, diseases, and mammal damage, while stressing the importance of monitoring and adaptive management. Biodiversity conservation efforts were discussed, with an emphasis on maintaining old-growth forest characteristics and mitigating the impact of invasive species. Additionally, the section examined ecosystem services provided by the UTHF, including carbon sequestration, water supply, and cultural value.

Throughout the book, the authors emphasized the necessity of adaptive and innovative forest management approaches in the face of global change. We advocated for long-term monitoring and research to better understand complex ecosystem dynamics and inform sustainable decision-making. Integrating ecological, social, and cultural goals was highlighted as essential for ensuring the resilience and health of forest ecosystems over the long term.

In this concluding chapter, we summarize challenges and opportunities for future research at the UTHF, particularly in the fields of ecology, forest management, and the provision of ecosystem benefits. By presenting the SSMS as a model for integrative forest management and silviculture, we hope to inspire its application in forest ecosystems around the world.

17.2 Future Challenges and Opportunities

17.2.1 Integrating Technological Advancements

Advancements in technology, such as remote sensing, geographic information systems (GIS), laser scanning, and unmanned aerial vehicles (UAVs), are revolutionizing forest management by improving efficiency, data collection, and decision-making (Chaps. 2, 8, and 16). These tools enable precise forest type classification, inventory, and harvesting operations, as demonstrated by their use in the UTHF, where technologies like laser scanning, UAVs, and electronic field books have enhanced management practices.

Effectively integrating these technologies into silvicultural practices supports sustainable forest management by optimizing operations and addressing complex challenges. However, their adoption must consider ethical and practical concerns, including privacy, data security, and equitable access to benefits. Continued investment in research and development will be crucial for advancing these tools and ensuring their responsible use in managing and conserving multiple ecosystem functions.

17.2.2 Ensuring Successful Natural Regeneration

Natural regeneration is a key aspect of integrative forest management in northern Japan. However, it faces challenges due to various factors such as dense dwarf bamboo, which restricts tree seedling establishment (Chaps. 7, 8, 10, and 12). *Picea jezoensis*, an important native conifer species in the UTHF, relies heavily on fallen logs for seedling establishment, making it particularly vulnerable to silvicultural management practices that prevent the supply of fallen trees or alter stand structure unfavorably for *P. jesozensis* regeneration (Chaps. 8, 12, and 14). To ensure successful regeneration for this species, careful harvesting practices, nurse log management, and silvicultural techniques like soil scarification and supplemental planting are required. In response to the declining number of suitable *P. jezoensis* seed trees, seed orchards have been established (Chap. 11). Additionally, experiments with soil scarification after group selection cutting aim to create suitable regeneration sites.

A rare, large-scale mass flowering and subsequent dieback of dwarf bamboo, *Sasa senanensis*, occurred throughout the Hokkaido Island between 2022 and 2023, marking the first such event in 120 years. *S. senanensis*, which densely covers the forest floor, typically restricts tree regeneration (Chap. 5). Its dieback could significantly enhance tree regeneration in previously poorly regenerated forest types, such as "Conifer-Dominant Selection Cutting Forests with Poor Regeneration," "Broadleaf-Dominant Selection Cutting Forests," and "Sparse Forests" (Chap. 8). These changes may transform these areas into sites with rich regeneration, influencing long-term forest dynamics, management strategies, and timber production. To study these impacts, long-term monitoring has been initiated in the UTHF (Fig. 17.1).

Fig. 17.1 Mass die-back of *Sasa senanensis* in 2023. (Photo by Satoshi N. Suzuki, 28 May 2024)

17.2.3 Maintaining Genetic Diversity

Conserving the genetic diversity of forest tree populations is essential for their long-term adaptability and resilience to environmental changes (Chap. 11). Forest management practices, such as selection harvesting and the use of limited seed sources, can unintentionally reduce genetic diversity (Chaps. 7 and 11). To address this, careful planning is needed to ensure diverse seed sources, sustainable harvesting practices, and the preservation of varied tree populations within managed forests.

The UTHF has established clonal seed orchards to produce high-quality seedlings for reforestation efforts. However, it is critical that these orchards capture a broad range of genetic variation from the original forest populations to avoid genetic bottlenecks. Additionally, preserving natural regeneration processes and maintaining a diversity of tree ages and sizes in managed forests are key strategies for supporting genetic diversity.

17.2.4 Addressing Climate Change Impacts

Climate change is expected to increase the frequency and severity of disturbances such as windstorms, insect outbreaks, snow damage, and wildfires (Chaps. 10, 13, and 14). These disturbances pose significant risks to forest ecosystems and timber production, particularly in monoculture plantations, which are more vulnerable. For example, simulations in the UTHF predict reduced net CO_2 absorption due to more frequent and severe windthrow events under future climate scenarios (Chap. 16).

Promoting mixed-species forests over monocultures can enhance resistance to wind and bark beetle disturbances, contributing to forest resilience (Chap. 10). Additionally, management practices such as leaving downed logs after windstorms can support natural regeneration, enhance carbon storage, and provide habitats for saproxylic organisms (Chaps. 12 and 16).

In hemi-boreal forests, a decrease in evergreen conifers and an increase in deciduous broadleaf species have been observed (Suzuki et al. 2015, Searle and Chen 2017, Hiura et al. 2019, Hisano et al. 2021, also refer to Chap. 5). This trend is likely influenced by climate changes, including rising temperatures (Chap. 4), altered precipitation patterns, reduced snowfall, and earlier snowmelt, as well as an increase in extreme climatic events like heavy rainfall, megadroughts, and super typhoons. These changes affect tree growth, mortality, regeneration, and the dynamics, structure, and composition of forest stands. Consequently, forest management strategies must be adapted to mitigate these effects and promote ecosystem resilience.

Alpine species are particularly threatened by climate change due to habitat reduction caused by rising temperatures. For instance, the Japanese pika (*Ochotona hyperborea*), an alpine species in the UTHF, faces survival risks as its habitat shrinks (Chap. 15). These challenges highlight the need for adaptive forest management practices that conserve biodiversity and maintain ecosystem functions in the face of changing environmental conditions.

17.2.5 Managing Forest Pests and Diseases

Managing insect pests and diseases is essential for sustainable forest management. Research into the mechanisms behind insect outbreaks and disease occurrences can help inform effective management strategies.

Defoliation of Monarch birch (*Betula maximowicziana*), a high-value timber species, by pests such as the Saturn moth (*Caligula japonica*) can trigger secondary attacks by ambrosia beetles (Chap. 14), leading to significant economic losses. Controlling these pests is crucial for the sustainable production of high-value timber. The attack of bark beetles on spruce species following disturbances or logging must also be taken into account (Chaps. 12 and 14). Natural regeneration of spruce species is challenging in selection logging stands (Chap. 12), making the management of bark beetles essential for maintaining spruce resources. Moving forward, it is crucial to monitor the impacts of climate change and the increase in disturbances it causes on bark beetle populations.

Oak wilt, a fungal disease vectored by the Japanese oak ambrosia beetle, *Platypus quercivorus*, has spread across Japan and was first confirmed in the southernmost region of Hokkaido in 2023, with 15 oak trees reported as affected (Hokkaido Prefecture 2024). By 2024, the reported number of affected trees had increased to 182, raising concerns about further expansion in the future. At present, the likelihood of an outbreak occurring in Furano may be low due to the region's cooler climate. However, with ongoing climate warming, this risk is expected to increase. It is therefore essential to consider and prepare measures to address this potential threat.

Wood decay is more prevalent in natural forests than in plantations, posing challenges for maintaining timber quality and forest health (Chap. 14). Determining the optimal cutting age for different tree species is critical to minimizing timber loss due to decay. Additionally, there is a need to develop reliable methods for diagnosing wood decay inside trees, reducing reliance on the expertise of experienced foresters.

17.2.6 Mitigating the Impact of Deer Overpopulation

Increasing deer populations pose a significant threat to forest ecosystems due to excessive feeding damage (Chaps. 12 and 15). Deer is often attracted to open areas created by forest management practices, such as clear-cuts and sparse forests. Their browsing on regenerating trees can hinder forest recovery and alter forest structure, particularly in areas with high deer densities. Overgrazing by deer not only prevents the establishment of tree seedlings but also reduces plant diversity and favors certain species over others. Additionally, bark stripping in sparse forests with dense dwarf bamboo further accelerates forest degradation.

To address these challenges, forest managers must develop effective strategies to control deer populations and mitigate their impacts. This includes careful consideration of harvesting practices and minimizing the creation of open areas that attract deer.

17.2.7 Controlling and Managing Invasive Species

The growing presence of nonnative species, including both plants and animals, poses a significant threat to forest ecosystems by disrupting native biodiversity and ecological functions (Chap. 15). Invasive species can outcompete native plants, alter food webs, introduce diseases, and modify habitat structure. For instance, black locust (*Robinia pseudoacacia*), an invasive plant species in the UTHF, outcompetes native trees in river side and forest edge, and has the potential to spread into open areas within the forest interior. Efforts to control its spread are essential.

Raccoons (*Procyon lotor*), another nonnative species observed in the UTHF, may impact the forest ecosystem, though their specific effects remain unclear. The introduction of exotic tree species for plantations also raises concerns about potential invasiveness (Chap. 10). Monitoring and managing the spread of such species is necessary to mitigate their impact on native ecosystems.

Preventing the introduction and spread of invasive species, along with managing existing populations, will be crucial to protecting forest ecosystems and maintaining their biodiversity and functionality.

17.3 Towards Harmonizing Conservation and Production Through Integrative Forest Management

Forests provide a wide range of ecosystem services, including timber production, carbon storage, water regulation, biodiversity conservation, and cultural services (Chap. 16). Balancing these services is essential for sustainable forest management, as prioritizing one service, such as intensive timber harvesting, can negatively impact others, including carbon storage, water flow, and biodiversity (Chap. 12).

Achieving this balance is a key challenge. The SSMS implemented at the UTHF seeks to address this by promoting natural regeneration, maintaining diverse forest structures, and avoiding intensive operations (Chap. 7). However, even under the SSMS, continuous evaluation and adaptive management are necessary to ensure the long-term sustainability of multiple ecosystem functions.

This book emphasizes the importance of developing sustainable forest management practices that balance timber production with biodiversity conservation and the provision of ecosystem services (Chaps. 7, 12, and 16). Addressing future challenges requires a holistic and adaptive approach that recognizes the interconnectedness of ecosystem functions. Long-term monitoring and research are essential for understanding the complex interactions within forest ecosystems and for informing effective management strategies. Collaboration with stakeholders is also crucial to ensure the development and implementation of innovative and sustainable solutions.

References

Hiura T, Sato G, Iijima H (2019) Long-term forest dynamics in response to climate change in northern mixed forests in Japan: a 38-year individual-based approach. For Ecol Manag 449:117469

Hisano M, Ryo M, Chen X, Chen HY (2021) Rapid functional shifts across high latitude forests over the last 65 years. Glob Chang Biol 27:3846–3858

Hokkaido Prefecture (2024) Oak wilt damage. https://www.pref.hokkaido.lg.jp/sr/srs/new23/206717.html. Accessed 26 Dec 2024

Searle EB, Chen HY (2017) Persistent and pervasive compositional shifts of western boreal forest plots in Canada. Glob Chang Biol 23:857–866

Suzuki SN, Ishihara M, Hidaka A (2015) Regional-scale directional changes in abundance of tree species along a temperature gradient in Japan. Glob Chang Biol 21:3436–3444

MIX
Papier aus verantwortungsvollen Quellen
Paper from responsible sources
FSC® C105338

If you have any concerns about our products,
you can contact us on
ProductSafety@springernature.com

In case Publisher is established outside the EU,
the EU authorized representative is:
Springer Nature Customer Service Center GmbH
Europaplatz 3, 69115 Heidelberg, Germany

Printed by Libri Plureos GmbH
in Hamburg, Germany